Bacterial Genomics

Genome Organization and Gene Expression Tools

Aswin Sai Narain Seshasayee

CAMBRIDGE
UNIVERSITY PRESS

Cambridge House, 4381/4 Ansari Road, Daryaganj, Delhi 110002, India

Cambridge University Press is part of the University of Cambridge.

It furthers the University's mission by disseminating knowledge in the pursuit of education, learning and research at the highest international levels of excellence.

www.cambridge.org
Information on this title: www.cambridge.org/9781107079830

First published 2015

Printed in India by Shree Maitrey Printech Pvt. Ltd., Noida

A catalogue record for this publication is available from the British Library

Library of Congress Cataloging-in-Publication Data
Seshasayee, AswinSaiNarain, author.
Bacterial genomics : genome organization and gene expression tools / AswinSaiNarainSeshasayee.
p. ; cm.
Includes bibliographical references and index.
Summary: "Discusses the application of genomic tools in the study of bacterial adaptation and provides review of recent research in the field of bacterial research"--Provided by publisher.
ISBN 978-1-107-07983-0 (hardback)
I. Title.
[DNLM: 1. Bacteria--genetics. 2. Genome, Bacterial--genetics. 3. Gene Expression Regulation, Bacterial. 4. Gene Expression. 5. Models, Genetic. QW 51]
QH434
572.8'6293--dc23
2014027746

ISBN 978-1-107-07983-0 Hardback

To

National Centre for Biological Sciences, Tata Institute of Fundamental Research, Bangalore, India

and

Dedicated to the memory of the ever-enchanting Professor K. S. Krishnan

Contents

List of Figures

Acknowledgements

I would like to thank my host institute, the National Centre for Biological Sciences (Bangalore, India), for providing me the resources and support to not only pursue research, but also to work on this book. All the members of my laboratory have been helpful and understanding during the times when I committed more hours to writing the book than to the lab. I would like to acknowledge the efforts of the two academic genomics core facilities that have enabled my research. These are the GeneCore facility at the European Molecular Biology Laboratory, Heidelberg, and the Next Generation Genomics Facility at the Centre for Cellular and Molecular Platforms, Bangalore. Thanks to the Department of Science and Technology (their research grants and the Ramanujan Fellowship scheme) and the Department of Biotechnology, Government of India, and CEFIPRA for funding our work.

Special thanks to Avantika Lal in the laboratory for designing and making several illustrations in this book. Aalap Mogre and Vittore Scolari from the lab also made illustrations for this work. Thanks to Hardik Gala (Institute of Stem Cell Biology and Regenerative Medicine, Bangalore) for contributing an illustration from his unpublished data. Many thanks to the large number of scientists worldwide who have published their high-quality work in open access journals, permitting free reproduction of important figures, which I could have never made otherwise. These, and others who gave permission to reuse copyrighted illustrations or data, are all acknowledged in the appropriate locations in the text. Of course, thanks to the entire community of researchers working in genomics and bacteriology!

Dasaradhi Palakodeti, Madan Babu Mohan, Subhajyoti De and Marco Cosentino Lagomarsino gave feedback on parts of the manuscript. This work would not have been possible without the efforts of Manish Chaudhari, my editor at CUP India.

Big thanks to Nicholas Luscombe, my PhD supervisor, for every great thing that a fantastic PhD supervisor does to his ward, in particular, give unbridled freedom to pursue one's interests in research.

My twin daughters were three months old when I started working on this book. They are now over two years old. A lot of credit to their impeccable behaviour and my wife Gayathri's efforts, which ensured that my work was not particularly affected by what was happening at home! Thanks, as always, to my parents for always being there.

Introduction: Bacterial Genomes and Gene Expression

1

Bacteria are the most dominant form of free-living life on Earth, and represent a major part of its genetic diversity. Some bacteria are closely adapted to a single environment in which they reside, whereas many others are capable of thriving across multiple environments. They range from the most benign inhabitants of the Earth to being deadly, multi-drug-resistant human pathogens. The success of bacterial life is exemplified by the wide variation in their genetic content, not just across phyla, but even among members of the same species. In addition is the plethora of gene regulatory mechanisms that—to state a most cliched phrase—ensure that appropriate genes are expressed when required. These two features of bacterial biology form the crux of this book.

It will not be wrong to state that the advent of genomics has considerably advanced our knowledge of both evolutionary and gene regulatory mechanisms in bacteria. The ambitious—and now historical—genome sequencing projects, driven by whole genome shotgun sequencing using automated Sanger sequencers, taught us many things about bacterial genetic diversity and the mechanisms underlying its generation. For example, by sequencing the genomes of many members of a species, such as *Escherichia coli*, we learnt that some bacteria have what is called an *open pan genome*: every new genome of a member of that species will invariably identify many genes hitherto unknown in that species. Genome sequencing projects also stoked a controversy on the importance of horizontal gene transfer to genome growth and bacterial evolution itself: in the face of rampant horizontal transfer, does the concept of a bacterial species have any meaning? At the other end of the scale, genome reduction emerged as an important phenomenon underlying the evolution of obligate parasites, including those of major human pathogens such as the *Rickettsia* and *Mycobacterium leprae*.

As databases of genome sequences grew exponentially in size, high-density microarrays, which use hybridisation-based techniques to probe nucleic acid content on a genomic scale, came into vogue. Though these were best known for their application to the measurement of gene expression on a genome-wide scale, these were also used extensively to probe the genetic content of large numbers of bacterial isolates for which the complete genome sequence of a close relative was available. Such studies, besides cataloguing gene signatures of specific isolates of a pathogenic species, also enabled studies of the dynamics of gene gain and loss. In the more classical application of studying gene expression, microarrays—either from a single study or through metaanalyses of data from multiple studies—enabled the reconstruction of gene regulatory networks of model bacteria, revealing a much under-appreciated complexity in the transcriptional response of even some so-called simple bacteria.

Arguably, the most dramatic advance in genomics came about recently with the advent of what are known as 'next-generation' or 'deep' sequencing technologies. These dramatically reduced the cost and the time required for sequencing complete genomes, thus bringing genome sequencing out of the confines of large sequencing centres to the benchtop of common laboratories. For example, the sequence of the *E. coli* strain, which was responsible for the break-out of food poisoning in Europe in 2011, was fully obtained on a benchtop sequencer within weeks after the outbreak, in stark contrast to the years of meticulous work that sequencing by the Sanger method required. Deep sequencing technologies have permitted high-resolution phylogenetic analysis of bacterial pathogens—including the tracking of epidemics—at times within clinically-relevant timescales. For the basic sciences researcher, these techniques have enabled rapid identification of genetic variants associated with a phenotype of interest. Genome sequencing, both classical and next-generation, have also spawned and advanced the field of metagenomics, which pertains to the genetic characterisation of entire bacterial communities in a culture-independent manner. A particular contribution of the next-generation sequencing approaches to this field has been in allowing us to obtain the complete or near-complete genome sequence of a single bacterial species from within a complex community of meta-genomes. One can even isolate a single bacterial cell from a consortium of bacteria and sequence its genome to a near-complete stage.

Finally, the great depth of coverage that many next-generation sequencers offer also makes them quantitative, allowing us to measure the extent to which a particular genetic variant has spread in a population, as well as pursue genome-scale assays of gene expression and protein–nucleic acid interactions, previously performed using microarrays. In contrast to microarray technologies, deep sequencing can provide base-level resolution in defining transcripts and protein-binding regions on DNA or RNA. This has considerably enhanced research into experimental annotation of transcripts and regulatory regions in genomes, an approach that has used both microarrays and deep sequencing technologies. Single-molecule real-time sequencing technologies also help identify base modifications in a standard sequencing experiment, thus significantly advancing our ability to describe epigenetic modifications.

When it comes to next-generation or deep sequencing, the sky is the limit. Virtually any molecular technique that uses sequence-level information about a few genes can be ported to a genomic scale using these techniques. For example, one can rapidly quantify locus-dependent rates of transposon insertion on a genomic scale using deep sequencing techniques called TraDIS and InSEQ, among others.

The objective of this text is to be a celebration of the application of genomics to the study of genome architecture and gene expression in bacteria. The book takes the approach of introducing concepts underlying the generation and analysis of multiple types of genome-scale data, followed by the presentation of a few papers which have either pioneered the application of a technology to a particular problem, and/or described interesting biology using a combination of genomic techniques. In the context of data analysis, we have deliberately stayed away from providing tables of software that can be applied to a particular type of data. While mentioning a few popular software in the main text, we have emphasised more on the fundamental principles that underly some of these software. Lists of relevant software are generally available online, as well as in the reviews that we cite in this text. Even in a relatively young field like genomics, it is impossible to be comprehensive in a book of this size. Therefore, the selection of concepts and case studies is merely a reflection of my own biases, interests and familiarity towards a subset of the fast-growing literature in the field. In fact, there are whole areas of genomics that this book does not touch upon—proteomics and metabolomics, for example.

Technology moves forward faster than many of us can cope with; therefore, the power of the Internet. However, I do believe that many ideas discussed in this text, published in the traditional format, whether a part of history or contemporary, will remain relevant for a long time to come.

2 Comparative Genomics in the Era of Sanger Sequencing

2.1 Introduction

The first genome to be completely sequenced,[1] in 1976, was that of an RNA bacteriophage MS2, which encoded only three genes. Though the concept of comparative genomics had yet to emerge, the authors mentioned the possibility of understanding viral evolution by performing sequence searches against other viral genomes that were forthcoming. This was rapidly followed by the sequencing–by Sanger and colleagues–of the 5.4 kb-long genomic DNA of phage ΦX174,[2] and later in 1982, considerably larger sequence (>48 kb) of the genome of the molecular biology workhorse, bacteriophage lambda (λ).[3] As pointed out by Koonin and Galperin,[4] neither work presented the concept of sequence comparison and homology, which is now routine in any genome analysis pipeline. This may be surprising as the PIR protein sequence database[5] was already available, and the first sequence substitution matrix[6] constructed. However, sequence databases

---

1 The sequence of the third and final gene was reported in Fiers W., Contreras R., Duerinck F., Haegeman G., Iserentant D., Merregaert J., Min Jou W., Molemans F., Raeymaekers A., Van den Berghe A., Volckaert G. and Ysebaert M. 1976. 'Complete nucleotide sequence of bacteriophage MS2 RNA: Primary and secondary structure of the replicase gene.' *Nature* 260: 500–07.

2 Sanger F., Coulson A. R., Friedmann T., Air G. M., Barrell B. G., Brown N. L., Fiddes J. C., Hutchison C. A. 3rd, Slocombe P. M. and Smith M. 1977. 'Nucleotide sequence of bacteriophage ΦX174 DNA.' *Nature* 265: 687–95.

3 Sanger F, Coulson A. R., Hong G. F., Hill D. F. and Petersen G. B. 'Nucleotide sequence of bacteriophage lambda DNA.' *Journal of Molecular Biology* 162: 729–73.

4 Koonin and Galperin. 2003. *Sequence-Evolution-Function: Computational Approaches in Comparative Genomics.* Kluwer Academic Press, Boston.

5 http://pir.georgetown.edu/

6 Dayhoff, Schwartz and Orcutt. 1978. 'A model for evolutionary change in proteins.' *Atlas of Protein Sequence and Structure* 5: 345–52.

were not sufficiently rich in information at that time, and sophisticated and rapid methods for searching sequence databases were not common, with the Smith–Waterman algorithm a then-recent innovation,[7] and the much faster BLAST not due to be reported for another eight years.[8]

The first tentative comparative genomics work, as identified by Koonin and Galperin, was published by Toh, Hayashida and Miyata in 1983.[9] This remarkable short paper identified homologues of retroviral reverse transcriptases encoded by the genome of two DNA viruses, whose life cycle was known to involve reverse transcription. In another study, Argos and coworkers[10] showed striking similarities in certain gene sequences between animal picornaviruses and plant cowpea mosaic virus, suggesting evolutionary relatedness either by independent abstraction of common host genes, or via vertical descent from a common ancestor. Finally, McGeoch and Davison[11] worked on considerably larger genomes of the varicella zoster virus (an α-herpesvirus), and the Epstein–Barr virus (a distantly related γ-herpesvirus), and found several homologues between them.

Despite several reports describing and comparing viral genomes, genomics of cellular organisms did not come to fruition till the publication of the genome of the human pathogenic bacterium *Haemophilus influenzae* in 1995,[12] almost 20 years after the completion of the genome of the bacteriophage MS2. Today, we have access to over 3,000 fully-sequenced bacterial genomes and many more draft genomes.

---

7 Smith and Waterman. 1981. 'Identification of common molecular subsequences.' *Journal of Molecular Biology* 147: 195–97.

8 Altschul S. F., Gish W., Miller W., Myers E. W. and Lipman D. J. 1990. 'Basic local alignment search tool.' *Journal of Molecular Biology* 215: 403–10.

9 Toh, Hayashida and Miyata. 1983. 'Sequence homology between retroviral reverse transcriptase and putative polymerases of hepatitis B virus and cauliflower mosaic virus.' *Nature* 305: 827–29.

10 Argos P., Kamer G., Nicklin M. J. and Wimmer E. 1984. 'Similarity in gene organization and homology between proteins of animal picornaviruses and a plant comovirus suggest common ancestry of these virus families.' *Nucleic Acids Research* 12: 7251–7267.

11 McGeoch and Davison. 1986. 'DNA sequence of the herpes simplex virus type 1 gene encoding glycoprotein gH, and identification of homologues in the genomes of varicella-zoster virus and Epstein-Barr virus.' *Nucleic Acids Research* 14: 4281–4292.

12 Fleischmann, R. D., Adams, M. D., White, O., Clayton, R. A., Kirkness, E. F., Kerlavage, A. R., Bult, C. J., Tomb, J., Dougherty, B. A., Merrick, J. M., McKenney, K., Sutton, G. G., FitzHugh, W., Fields, C. A., Gocayne, J. D., Scott, J. D., Shirley, R., Liu, L. I., Glodek, A., Kelley, J. M., Weidman, J. F., Phillips, C. A., Spriggs, T., Hedblom, E., Cotton, M. D., Utterback, T., Hanna, M. C., Nguyen, D. T., Saudek, D. M., Brandon, R. C., Fine, L. D., Fritchman, J. L., Fuhrmann, J. L., Geoghagen, N. S., Gnehm, C. L., McDonald, L. A., Small, K. V., Fraser, C. M., Smith, H. O. and Venter, J. C. 1995. 'Whoge-genome random sequencing and assembly of Haemophilus influenzae Rd.' *Science* 269: 496–98 + 507–12.

In this chapter, we provide a conceptual overview of the process of bacterial genome sequencing and annotation. This will be followed by a description of selected bacterial genomes and the lessons learnt from them. We will close the chapter by discussing some general trends and concepts that have emerged from large-scale comparative genomics of bacterial (and where appropriate, archaeal and eukaryotic) genomes.

2.2 The process of assembling and annotating bacterial genomes

DNA sequencing produces 'reads'–sequences of fragments of DNA of orders of magnitude shorter than the genome itself. A typical 500 nt read from a Sanger sequencing experiment is ~10^4-fold smaller than the >4 Mb genome of the model laboratory strain of *Escherichia coli*. Therefore, novel DNA fragmentation strategies had to be developed so that sequencing reads produced from such fragments could be *assembled* into a complete chromosome(s). The most straightforward procedure for generating fragments, without recourse to any genetic or physical mapping, is random shearing of the genomic DNA, called the whole genome shotgun method (Fig. 2.1; top). It follows from the theory of Lander and Waterman[13] that, given a sufficiently random fragment set, the probability that any nucleotide position in the genome is covered by k reads is approximated by a Poisson distribution, with deviations caused by biological factors such as clone lethality. On the basis of these calculations, for a sequencing coverage of m, the probability that a base is not sequenced is given by $P = e^{-m}$; this number equals 0.0067 for $m = 5$, and 0.000045 for $m = 10$.

Despite the simplicity and force of this theoretical argument, there were justifiable fears that the scale of data produced might be too complex to handle during assembly. These fears were laid to rest by the publication of the genome of *H. influenzae*, a genome obtained *de novo* without any reference maps. Besides establishing the validity of the whole genome shotgun procedure, the work also laid down a data analysis pipeline, which includes read assembly, gap closure, gene finding and gene annotation, in that order. We will now briefly look at each of these steps (Fig. 2.1).

---

13 Lander and Waterman. 1988. 'Genomic mapping by fingerprinting random clones: A mathematical analysis.' *Genomics* 2: 231–39. Described in simple terms in the *H. influenzae* genome paper.

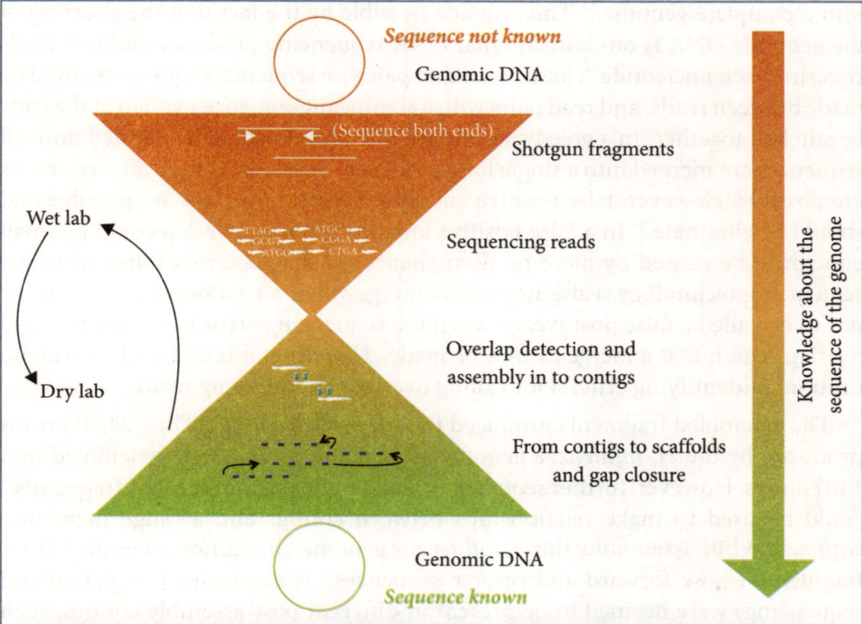

Figure 2.1 *A schematic overview of the process of sequencing a bacterial genome.* The top cone represents the fragmentation of genomic DNA molecules to produce *shotgun fragments*, the ends of which are sequenced. Typically, both ends of each fragment are sequenced, producing what are known as *paired-end* reads. The bottom cone shows the computation-intensive part of a genome sequencing experiment where sequence reads are assembled to produce larger *contigs* (and *scaffolds*, in which adjacent contigs are linked together); at this stage, there are un-resolved gaps between contigs/scaffolds, and the linear order of contigs is not necessarily known. A combination of computation and laboratory experiments is then used to stitch contigs together in a process called *gap closure*. The unique advantage of paired-end reads during the assembly of reads into a genome is mentioned in Chapter 4. The width of the cone is also indicative (not-to-scale) of the size of the average DNA molecule being analysed at any given stage of the experiment.

2.2.1 Genome assembly and gap closure

In a typical genome sequencing effort, either ends (*paired-end sequencing*) of DNA fragments of length ~2 kb are sequenced to produce *reads*, which in Sanger sequencing are typically ~500 nt long; for example, sequencing the genome of *H. influenzae* produced ~24,000 fragments of an average length of ~450 nt. The first step, after the production of these DNA sequence reads, is to assemble the reads together into longer fragments called *contigs* and *scaffolds*, and eventually into a complete genome.[14] This is made possible by the fact that the shearing of the genomic DNA is random and that DNA sequencing produces multiple reads covering each nucleotide. These mean that pairwise sequence alignments could be made between reads, and read pairs with a significant sequence overlap at the ends be stitched together. In a greedy algorithm, this process could be iterated until all sequences are merged into a single long sequence, or at least into as few sequences are possible. However, false positive and false negative overlaps are possible and should be eliminated. In a false positive identification, an overlap could be small enough to be caused by mere random chance, or due to errors inherent to the sequencing technology. False negatives emerge when a true overlap is too short not to be called a false positive, or when the sequencing error is so high in a real overlap region that a merge cannot be made. Therefore, it is essential to exercise caution in identifying criteria for calling overlaps and merging reads.

The assembled fragments produced thus are called *contigs*. The ~24,000 reads produced by the *H. influenzae* genome sequencing project were assembled into 140 contigs. However, further sequencing reactions using longer DNA fragments[15] could be used to make relationships between contigs and arrange them into *scaffolds*. While assembling the *H. influenzae* genome, the authors identified DNA fragments whose forward and reverse sequences (as determined by paired-end sequencing) were deemed to be present in different post-assembly contigs; such contig pairs were linked together and eventually, the 140 contigs were organised into 42 groups separated by gaps. These gaps could be finally closed using PCR reactions primed by sequences from the edges of each pair of scaffolds, followed by the sequencing of appropriate amplicons. These were helped by analyses such as searches against protein databases, wherein two contigs that find a match against two different portions of a single protein sequence could be tentatively deemed to lie adjacent to each other. Finally, gaps that could not be closed because of lethality resulting from the cloning of certain DNA fragments in the host *Escherichia coli* could be sequenced using a phage library; in fact, 23 of the 42 physical gaps in the *H. influenzae* genome were closed using this approach.

2.2.2 Genome-scale computational identification of features

Once the complete genome sequence of an organism is obtained, the next step is to identify various functional features. First, the genomic G+C content is described. This simple measure is useful as it could be an overall indicator of the various mutational pressures operating on a genome.[16] It also allows a first-pass identification of genes with a different evolutionary history from the rest

---

14 Sutton and Dew. 2006. 'Shotgun fragment assembly.' *Systems Biology Volume 1: Genomics*. Oxford University Press, USA.

15 See reference to mate-pair library sequencing in Chapter 4.

16 See discussion on genome reduction later in this Chapter.

of the genome, as many of these lie in windows whose G + C content is different from the genomic average (Fig. 2.2 a). Similarly, ribosomal operons, which are highly conserved, also tend to lie within regions of high G + C content. Patterns of G/C usage are also useful in identifying the origin of replication. Since the leading and the lagging strand are replicated differently, and may therefore come under different mutational regimes, they have different G/C skews as calculated by G – C / G + C. This is called strand asymmetry. Any strand of DNA will be lagging on one side of the origin and leading on the other; therefore, a position around which there is a strong switch in the G/C-skew value is a candidate for being the origin of replication (Fig. 2.2 b). This measure is usually supplemented

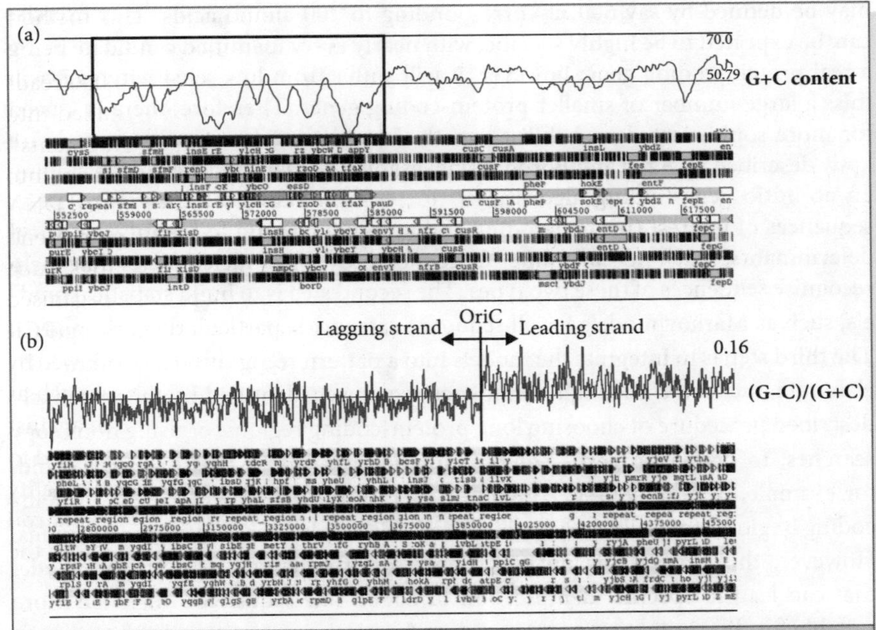

Figure 2.2 *Base composition patterns.* (a) The G+C content over short windows for a segment of the genome of *Escherichia coli* K12 is shown. The boxed region is one where the G+C content is considerably less than the average, which is marked by the thin black horizontal line as well as by the blue curve to the right of the box. This region of abnormally low A+T content is a horizontally-acquired piece of DNA, from a defective prophage called DLP-12. (b) The G/C-skew across a portion of the *E. coli* K12 genome is shown. The switch in the skew from negative (lagging strand) to positive (leading strand) at the origin of replication is marked.

by other predictors such as the presence of clusters of the binding site for the DNA replication initiator protein DnaA. The terminus is usually bipartite with the two parts bisected by a position that is diametrically opposite to the origin of replication on the circular chromosome.

Next, highly conserved genes such as the ribosomal and transfer RNAs can typically be identified by homology to known examples.

Today, though a large number of protein coding sequences can be identified by homology to members of rich sequence databases, the possibility of missing novel proteins still remains large. Therefore, homology-independent, but pattern-dependent statistical methods are first used to identify regions that might potentially encode proteins, before these are annotated with putative functions, typically by homology. A simple pattern defining a protein-coding region may be any 'long' stretch of sequence between a START and a STOP codon, where 'long' may be defined by say 600 nt corresponding to 200 amino acids. This method can be expected to be highly specific, with nearly every identified candidate being a real protein-coding gene; however, it will suffer from low sensitivity as it will miss a large number of smaller protein-coding genes. Therefore, there is a need for more sophisticated probabilistic methods for gene identification, which are aptly described by Azad and Borodovsky[17] as follows: "The first step in developing an ab-initio gene-finding algorithm is to perform statistical analysis of DNA sequences of interest (protein-coding and non-coding) and to identify statistical determinants, such as in-frame frequencies of oligonucleotides, that help recognise sequences of these two types. The second step is to build statistical models, such as Markov models for all sequence categories, particularly gene models. The third step is to integrate the models into a pattern recognition algorithm". To elaborate, one could use high-confidence sequences identified by the previously described procedure of choosing long protein-coding regions, or using homology searches, to define properties of protein-coding regions. This property could, for example, be simply G+C content; in fact, in the genome of *E. coli*, protein-coding regions typically have a higher G+C content than intergenic regions. However, this property is not sufficiently discriminative. A statistical model that can lead to specific identification of codon-like sequences might be more suitable for discriminating protein-coding from other sequences. In this respect, a base composition profile that is calculated for each position within a codon is a sensible predictor; for example, in typical *E. coli* genes, the base 'G' is twice as common at the first base of a codon as it is in the second base. However, such position dependence is not seen in intergenic regions. These occurrence profiles could be extended to include immediate sequence context in a Markov model,[18]

[17] Azad and Borodovsky. 2004. 'Probabilistic methods of identifying genes in prokaryotic genomes: Connections to the HMM theory.' *Briefings in Bioinformatics* 5: 118–30.

[18] An mth order Markov model in the context of a DNA sequence states that the probability of finding a particular base at position i is dependent only on m-preceding bases; a zeroth order Markov model is nothing beyond base composition, and a first order

i.e., the occurrence of a base at a particular codon position when preceded by the oligonucleotide X will be scored differently to when preceded by a different oligonucleotide Y, with the length of the oligonucleotide being the order of the Markov model, a parameter to be optimised. The presence of such patterns is then used to score any unannotated piece of genomic sequence; this score eventually determines whether a sequence is protein coding or not. A result of a toy Markov model, which shows that higher order Markov models differentiate between gene sequences and random sequences better than lower order models, is shown in Fig. 2.3. It also shows that a variable order Markov model, which takes the position

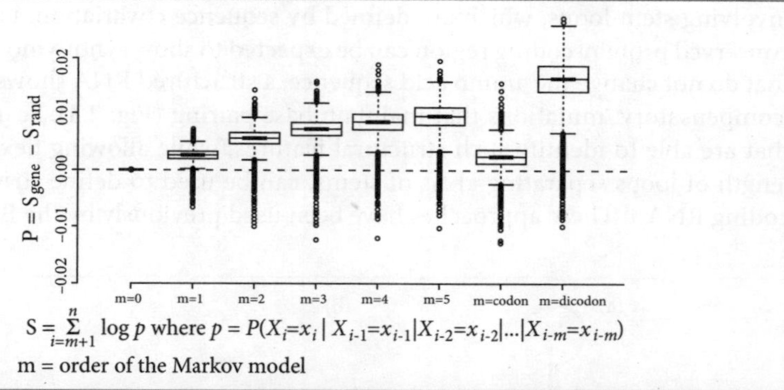

Figure 2.3 *A toy Markov model differentiating between genes and random sequences.* The likelihood that a random sequence resembles a real gene sequence in terms of its base composition, as defined by Markov models, decreases with increase in the order (m) of the model. The y-axis shows the difference in scores between a real gene sequence and a shuffled variant of the same sequence, given a Markov model of a certain m trained on a set of known gene sequences. This difference increases with m. For the codon and the dicodon models, m is not fixed and is determined by the position of a base in a codon or a pair of adjacent codons (for example, in the codon model, $m = 1$ for the second base of a codon and $m = 2$ for the third base). This is only a toy model, where the training and the test sequences were sourced from the same genome (*E. coli* K12); this is not necessarily reflective of real gene feature identification models.

of a base within the context of a codon pair, performs better than the corresponding fixed-order models. Gene identification programs currently in vogue,

Markov model can be derived simply from dinucleotide frequencies. Higher the order of the Markov model, better is the prediction likely to be (see Fig. 2.3). However, the use of higher order models is limited by practical difficulties such as obtaining sample sizes sufficient for a robust statistical analysis. These are discussed Salzberg S. L., Delcher A.L., Kasif S. and White O. 1998. 'Microbial gene identification by interpolated Markov models.' *Nucleic Acids Research* 26: 544–48.

such as GLIMMER[19] and GeneMark,[20] employ higher-order Markov models to achieve their objectives.

In recent times, various stable non-protein coding RNA–other than ribosomal and transfer RNA–have been discovered and annotated with important regulatory functions. An important class of such RNA involves riboswitches, which are present in the 5′-untranslated region of various mRNA, and have the ability to bind to small molecules. Similar to protein-coding regions, general patterns that describe architectural features of such RNA molecules can be used to detect them in complete genomes. These RNA are likely to be found only in long stretches of intergenic regions. They also have secondary structural motifs involving stem loops, which are defined by sequence covariation, i.e., whereas a conserved protein-coding region can be expected to show synonymous mutations that do not change the amino acid sequence, a structured RNA shows a pattern of 'compensatory' mutations that maintain base pairing (Fig. 2.4). Search methods that are able to identify such structural features, while allowing flexibility in the length of loops separating arms of stems, can be used to define structured non-coding RNA.[21] These approaches have been used previously by the Breaker lab to

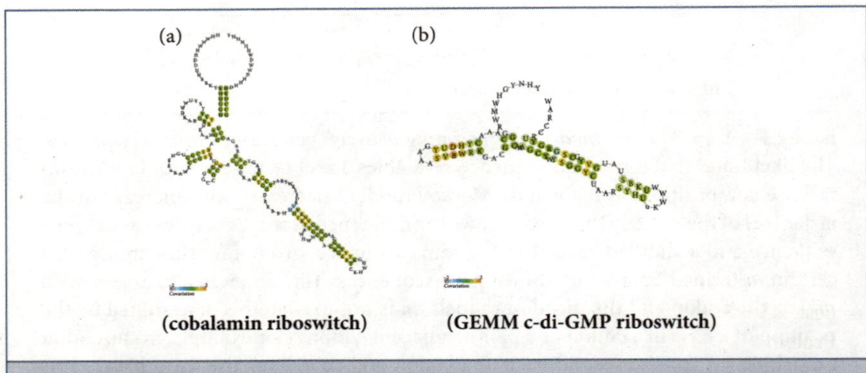

Figure 2.4 *Regulatory RNA secondary structures.* The secondary structures of riboswitches that bind (a) cobalamin and (b) cyclic-di-GMP. The coloured bases on the base-paired stems indicate high sequence covariation. These images were obtained from RFAM (rfam.sanger.ac.uk).

19 http://ccb.jhu.edu/software/glimmer/index.shtml, accessed on 9th July, 2014.

20 http://opal.biology.gatech.edu/, accessed on 9th July, 2014.

21 (a) Barrick J. E., Corbino K. A., Winkler W. C., Nahvi A., Mandal M., Collins J., Lee M., Roth A., Sudarsan N., Jona I., Wickiser J. K. and Breaker R. R. 2004. 'New RNA motifs suggest an expanded scope for riboswitches in bacterial genetic control.' *Proceedings of the National Academy of Sciences USA* 101: 6421–6426; (b) Rivas and Eddy. 2001. 'Noncoding RNA gene detection using sequence analysis.' *BMC Bioinformatics* 2: 8.

identify a large number of new riboswitch classes.[22] Sequence covariation data are available in a public database called RFAM,[23] which in conjunction with programs such as INFERNAL,[24] can be used to identify new examples of previously known structured RNA in unannotated genomes.

2.2.3 Annotating genes with functions

The next step in a genome project is the assignment of putative functions to genes (Fig. 2.5). This is generally done at the level of protein sequence by homology to previously sequenced proteins/genes, information about which

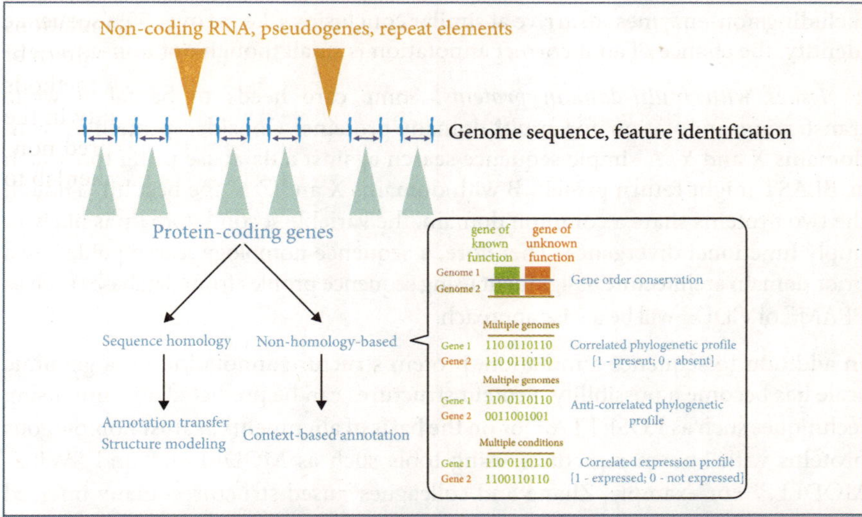

Figure 2.5 *A schematic outline of the process of annotating a bacterial genome.* Protein-coding features are annotated with function primarily by homology to other sequences in databases, and at times on a large scale by structure prediction and alignments. Genes can also be annotated based on guilt-by-association, where genes, which co-occur as part of a same operon, or be present or absent together across a variety of genomes, or be co-expressed over different environmental conditions, can be tentatively annotated as belonging to the same functional pathway. Intergenic sequences can encode regulatory RNA with defined secondary structures, which can be predicted using sequence covariation (see Fig. 2.4).

---

22 Weinberg Z., Wang J. X., Bogue J., Yang J., Corbino K., Moy R. H. and Breaker R. R. 2010. 'Comparative genomics reveals 104 candidate structured RNAs from bacteria, archaea, and their metagenomes.' *Genome Biology* 11: R31.

23 http://rfam.sanger.ac.uk, accessed on 9th July, 2014.

24 http://infernal.janelia.org, accessed on 9th July, 2014.

is available in public sequence databases. Typically, functional annotation is transferred from a homologue in the sequence database to the protein of interest. As straightforward as it looks, there are a few issues to be considered. These are as follows:

1 *What level of sequence similarity/identity is suitable for annotation transfer?* This issue has been addressed previously. Using enzyme commission (EC) numbers, Todd, Orengo and Thornton[25] showed that all four EC numbers tend to be the same for protein pairs with 40% sequence identity or more indicating a very high level of functional similarity; at a smaller sequence identity of 30%, two homologous proteins tend to share the first three EC numbers. Sangar and co-workers[26] used a broader definition of function based on gene ontology–thus including non-enzymes–to arrive at similar conclusions, i.e., at over 50% sequence identity, the chance of an incorrect annotation is small though not non-existent.

2 *Issues with multi-domain proteins.* Some care needs to be taken while transferring annotation for multi-domain proteins. Consider protein A with domains X and Y. A simple sequence search against a database using tools such as BLAST might return protein B with domains X and Z as the best hit. Though the two proteins share a common domain, the variable second domain is likely to imply functional divergence. Therefore, a sequence homology search guided by a prior domain architecture assignment using sequence profiles from databases such as PFAM[27] or CDD[28] will be a wise approach.

In addition to sequence annotation, protein structure annotation on a genomic scale has become a possibility. Protein structures can be predicted ab-initio using techniques such as ROSETTA,[29] or on the basis of alignments against homologous proteins with known structures using tools such as MODELER[30] and SWISS-MODEL.[31] For example, Zhang and colleagues[32] used structures–many inferred computationally–of nearly 500 metabolic enzymes encoded by the bacterium *Thermatoga maritima* to show that a small number of structural folds are

25 Todd, Orengo and Thornton. 2001. 'Evolution of function in protein superfamilies, from a structural perspective.' *Journal of Molecular Biology* 307: 1113–1143.

26 Sangar V., Blankenberg D. J., Altman N. and Lesk A. M. 2007. 'Quantitative sequence-function relationships in proteins based on gene ontology.' *BMC Bioinformatics* 8: 294.

27 http://pfam.sanger.ac.uk, accessed on 9th July, 2014.

28 http://www.ncbi.nlm.nih.gov/Structure/cdd/cdd.shtml, accessed on 9th July, 2014.

29 (a) http://www.rosettacommons.org/software/, accessed on 9th July, 2014; (b) http://robetta.bakerlab.org/, accessed on 9th July, 2014.

30 http://www.salilab.org/modeller/, accessed on 9th July, 2014.

31 http://swissmodel.expasy.org/, accessed on 9th July, 2014.

32 Zhang Y., Thiele I., Weekes D., Li Z., Jaroszewski L., Ginalski K., Deacon A. M., Wooley J., Lesley S. A., Wilson I. A., Palsson B., Osterman A. and Godzik A. 2009. 'Three-dimensional structural view of the central metabolic network of Thermotoga maritima.' *Science* 18: 1544–1549.

recruited to perform a variety of diverse yet related enzymatic functions. This is only in agreement with older observations–based on experimentally-determined protein structures from diverse organisms–that distinct substrate specificities and catalytic activities within a single protein superfamily can be accommodated by local mutations, surface loop variations and the use of additional domains.[33] Recently, protein structural information was used to predict protein–protein interactions on a genomic scale, albeit for the eukaryotes of yeast and human.[34] Depending on the quality of the structural model, fine-scaled applications such as protein-ligand docking for rational drug design can also be envisaged.

Though homology-based protein annotation, with or without information on protein structures, have been remarkably successful, many proteins–even in well-characterised organisms such as *E. coli*–remain 'hypothetical'. It is here that non-homology-based methods of annotating protein function, which are based on genomic context, become useful supplements (Fig. 2.5). These methods rely on our ability to perform effective comparative genomics across a variety of organisms, or our ability to integrate diverse functional genomic data[35] for a given organism. Some of these approaches are briefly described below, but have been reviewed extensively elsewhere.[36]

1 *Gene order conservation*: In the face of recombination, it is unlikely that the order in which genes are encoded on the chromosome is conserved across many species or genera. Therefore, such co-occurrence might in fact imply functional relatedness among the genes involved.

2 *Correlations between genes in their occurrence across multiple genomes*: If two genes show a tendency to be present or absent together, across a sufficiently large number of organisms, a functional link may be made between the two in the sense that they might act in the same pathway. If the occurrence of one excludes that of another, it might be indicative of functional replacement of one by the other, or antagonistic interactions between the two.

3 *Correlations in functional genomic data*: Where such data are available, the tendency of two genes to be co-expressed across many conditions, or be co-

---

[33] Todd, Orengo and Thornton. 2001. 'Evolution of function in protein super families, from a structural perspective.' *Journal of Molecular Biology* 307: 1113–1143.

[34] Zhang Q. C., Petrey D., Deng L., Qiang L., Shi Y., Thu C. A., Bisikirska B., Lefebvre C., Accili D., Hunter T., Maniatis T., Califano A. and Honig B. 2012. 'Structure-based prediction of protein-protein interactions on a genome-wide scale.' *Nature* 490: 556–60.

[35] Discussed in Chapter 5 in this book.

[36] (a) Aravind. 2000. 'Guilt by association: Contextual information in genome analysis.' *Genome Research* 10: 1074–1077; (b) Seshasayee and Madan Babu. 2005. 'Contextual inference of protein function.' *Encyclopedia of Genetics, Genomics, Proteomics and Bioinformatics,* edited by S. Subramanian. John Wiley and Sons, accessed on 9th July, 2014.

regulated by the same set of regulators, or be part of the same protein complex, might again indicate functional relatedness.

Large-scale functional annotations of genes in many genomes, on the basis of genomic context, is now available in the STRING database.[37]

2.3 Case studies

Having broadly described various components of a genome assembly and annotation pipeline, we will now briefly summarise the lessons learnt from a few early and iconic bacterial genome sequencing efforts. The selection of genomes and studies is subjective and might reflect this author's partiality towards certain concepts and studies. It is also emphasised here that these descriptions at best can only be brief–readers are referred to several books[38] that devote entire chapters for one class of organisms, and to the primary literature.

2.3.1 The *Escherichia coli* complex and large-scale horizontal gene acquisition

The bacterium *E. coli* holds an extraordinary position, not only as a model organism for molecular biology, but also as a major pathogen of humans and a crucial tool in biotechnology. It is an enterobacteriaceae (within γ-proteobacteria); a group which also includes other pathogenic genera like *Salmonella*, *Yersinia* and *Klebsiella*. The complete genome sequence of the non-pathogenic, laboratory-adapted strain of *E. coli,* K12 MG1655 (henceforth referred to as K12 in this section, or just as *E. coli*), was published after six years of painstaking work in 1997, by a group led by Frederick Blattner.[39] The ~4.6 Mb genome had nearly 50% G+C content, and was annotated with nearly 4,300 protein-coding genes (Fig. 2.6). These genes accounted for nearly 88% of the genome, leaving only 11% for non-genic regulatory and other functions, with a little over 1% covered by ribosomal and transfer RNA. A strong G/C skew switching was observed at the both the origin and the terminus

---

[37] http://string.embl.de/, accessed on 9th July, 2014.

[38] (a) Fraser, Read and Nelson. Eds. 2004. *Microbial Genomics*. Humana Press, USA. (b) Pallen, Nelson and Preston. Eds. 2007. *Bacterial Pathogenomics*. ASM Press, USA. (c) Baquero, Nombela, Cassell and Gutierrez-Fuentes. Eds. 2008. *Evolutionary Biology of Bacterial and Fungal Pathogens.* ASM Press, USA. (d) Hensel and Schmidt. Eds. 2008. *Horizontal Gene Transfer in the Evolution of Pathogenesis.* Cambridge University Press, USA.

[39] Blattner F. R., Plunkett G. 3rd, Bloch C. A., Perna N. T., Burland V., Riley M., Collado-Vides J., Glasner J. D., Rode C. K., Mayhew G. F., Gregor J., Davis N. W., Kirkpatrick H. A., Goeden M. A., Rose D. J., Mau B. and Shao Y. 1997. 'The complete genome sequence of *Escherichia coli* K-12.' *Science* 277: 1453–1462.

of replication (Fig. 2.2 b), thus ascribing differential mutational pressures to the asymmetry between the lagging and the leading strands in the manner in which they are replicated. The authors also calculated the codon adaptation index (CAI)–a measure of the tendency of a gene to use the most common codons in the genome–for all protein-coding genes and showed that many regions with low CAI corresponded to cryptic (or defective) prophages, and, in general, might be indicative of recent horizontal acquisition. As part of the annotation, the authors were able to detect new cryptic prophages, novel operons for the degradation of aromatic amino acids, and many genes for the flagellum that were shown to be nearly identical to the previously studied flagellar genes in *Salmonella*.

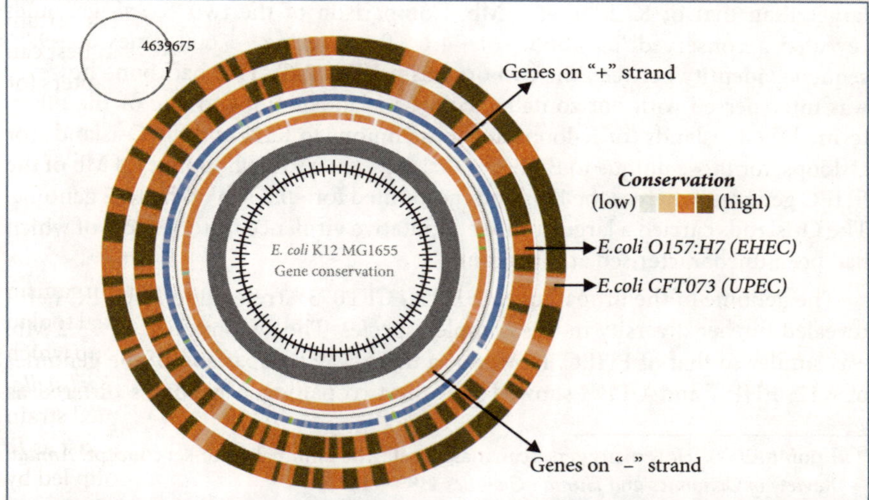

Figure 2.6 *The genome of E. coli K12 MG1655.* The genome of *E. coli* K12 MG1655 is shown. The two inner coloured bands mark the location of genes on the + and the – strand in blue and red respectively. The outer-most band marks the conservation of genes from *E. coli* K12 in the uropathogenic *E. coli* (strain CFT073), and the second outer band that in the enterohemorrhagic *E. coli* O157: H7. In the outer conservation bands, red and deep red indicate high levels of conservation, whereas white is indicative of a lack of conservation. Note the presence of several patches of white, identifying regions unique to *E. coli* K12. This figure was drawn using the GeneWiz browser online (http://www.cbs.dtu.dk/services/gwBrowser/).

A comparison of the genome of K12 with those of previously sequenced bacterial genomes–*H. influenzae, Mycoplasma genitalium* and *Synechocystis sp.*–showed that they had only a little over 100 genes in common. This analysis added to prior comparisons of *H. influenzae* and *M. genitalium* suggesting that the minimal genome of life could be much smaller than presumed.[40] Specifically, these comparisons showed that the number of genes involved in core processes such as translation was fairly similar across these diverse genomes. However, *E. coli* encoded a considerably larger number of genes involved in 'transport and binding' than the smaller genomes, indicating the greater metabolic flexibility of this organism.

A few years later, in 2001, the genome of the diarrhoeal-disease-causing enterohemorrhagic *E. coli* O157: H7 (referred to as EHEC in this section) was published by two groups, one from the USA[41] and the other from Japan.[42] This revealed a surprisingly large diversity in gene content between K12 and EHEC, two members of the same bacterial species. The genome of EHEC was much larger than that of K12, at ~5.6 Mb. Comparison of the two *E. coli* genomes revealed a conserved 'backbone' of ~4.1 Mb, covering ~3,600 genes at ~98% sequence identity between the two organisms (Fig. 2.6). This backbone however was interspersed with horizontally-acquired islands unique to one or the other, termed the K-islands (or K-loops, for those unique to K12) and the O-islands (or O-loops, for those unique to EHEC). O-islands covered a whopping 1.3 Mb of the EHEC genome, whereas the K-islands accounted for ~0.5 Mb of the K12 genome. The O-islands carried a large number of putative virulence factors, most of which had been uncharacterised at that point.

The genome of the uropathogenic *E. coli* CFT073[43] (refered to as UPEC here) revealed further diversity in this complex species. The genome size, at ~5.2 Mb, was similar to that of EHEC. However, a three-way comparison of the genomes of K12, EHEC and UPEC showed that the two pathogen genomes differed as

---

40 Koonin. 2000. 'How many genes can make a cell: The minimal-gene-set concept.' *Annual Review of Genomics and Human Genetics* 1: 99–116.

41 Perna N. T., Plunkett G. 3rd, Burland V., Mau B., Glasner J. D., Rose D. J., Mayhew G. F., Evans P. S., Gregor J., Kirkpatrick H. A., Pósfai G., Hackett J., Klink S., Boutin A., Shao Y., Miller L., Grotbeck E. J., Davis N. W., Lim A., Dimalanta E. T., Potamousis K. D., Apodaca J., Anantharaman T. S., Lin J., Yen G., Schwartz D. C., Welch R. A. and Blattner F. R. 2001. 'Genome sequence of enterohaemorrhagic *Escherichia coli* O157: H7.' *Nature* 409: 529–33.

42 Hayashi T., Makino K., Ohnishi M., Kurokawa K., Ishii K., Yokoyama K., Han C. G., Ohtsubo E., Nakayama K., Murata T., Tanaka M., Tobe T., Iida T., Takami H., Honda T., Sasakawa C., Ogasawara N., Yasunaga T., Kuhara S., Shiba T., Hattori M. and Shinagawa H. 2001. 'Complete genome sequence of enterohemorrhagic *Escherichia coli* O157: H7 and genomic comparison with a laboratory strain K-12.' *DNA Research* 8: 11–22.

43 Welch R. A., Burland V., Plunkett G. 3rd, Redford P., Roesch P., Rasko D., Buckles E. L., Liou S. R., Boutin A., Hackett J., Stroud D., Mayhew G. F., Rose D. J., Zhou S., Schwartz D. C., Perna N. T., Mobley H. L., Donnenberg M. S. and Blattner F. R. 2002. 'Extensive mosaic structure revealed by the complete genome sequence of uropathogenic *Escherichia coli*.' *Proceedings of the National Academy of Sciences USA* 99: 17020–17024.

much from each other as they did from the non-pathogenic strain. Nevertheless, it was reiterated that the variation was largely localised to various inserted islands, with the conserved backbone still existent but reduced to cover less than 3,100 genes. The UPEC strain lacked several virulence factors encoded by EHEC, leading to the interpretation that these differences could be responsible for UPEC's ability to harmlessly colonise the intestine. However, the large number of adhesive organelles in UPEC–again described later in the genome of another

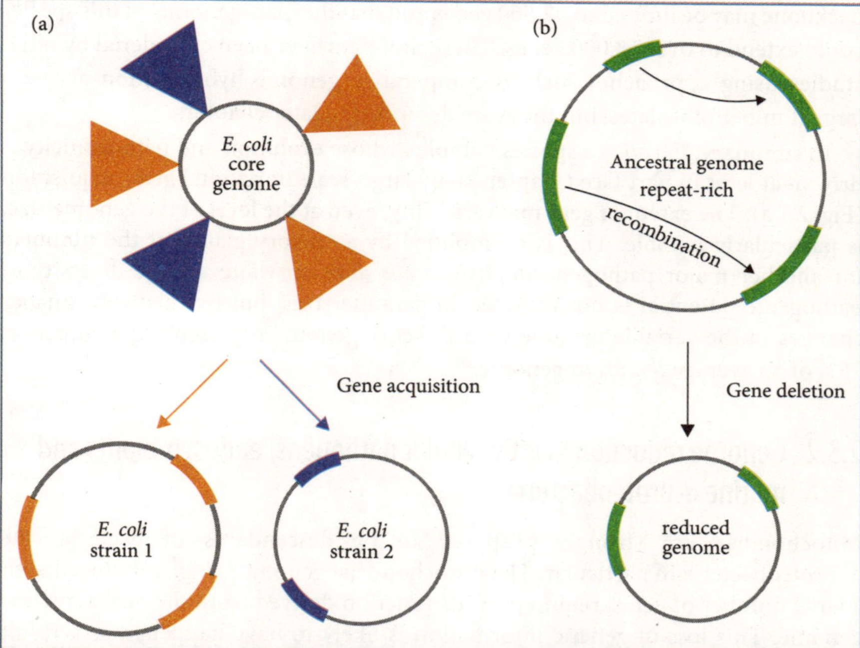

Figure 2.7 *The two major forces of large-scale bacterial evolution.* (a) Horizontal gene acquisition is a major force in the evolution of bacteria such as *E. coli*, and a contributor to genome growth. A relatively small core genome is conserved across members of the *E. coli* species complex, on to which isolate-specific gene islands are tagged on. Note that this is a simplistic view of *E. coli* genome evolution, which is driven probably by as much gene deletion as gene acquisition (see discussion of the evolution of *E. coli* gene content in Chapter 3). (b) Genome reduction is a common feature of obligate symbionts such as *Rickettsia*. Comparison of two *Rickettsial* genomes suggested that repeat-mediated homologous recombination is an important driver of genome reduction.

enterobacterial pathogen of the urinary tract, *Proteus mirabilis*[44]–might underlie specialisation for its niche.

Sequencing of several *E. coli* isolates across different pathovars revealed considerable genetic diversity among members of the same pathovar.[45] This study also identified genes, previously thought to be restricted to pathogens, in the genome of a commensal *E. coli* not known to cause disease even at high loads, leading to speculation on the role of these commensals in pathogen evolution. Finally, this study, by comparing 17 *E. coli* genomes, also showed that the common backbone may be limited to <2,500 genes and that the 'pan-genome' of this species could extend to over 13,000 genes. These numbers have been considered by other studies using approaches such as comparative genome hybridisation across a large number of isolates, but these are described in later chapters.

In summary, *E. coli* is a species complex whose evolution and pathogenicity is driven–at least in part (see Chapter 3)–by large-scale horizontal gene acquisition (Fig. 2.7 a). The extent of genomic variability, even at the level of the genome size, is particularly notable. This is exemplified by a cursory glance at the numbers for another major pathogen, *Staphylococcus aureus*, where a large diversity of pathogenic potential is not reflected in genome sizes, but by relatively smaller changes in the variable genome with the core genome representing as much as 75% of an average *S. aureus* genome.[46]

2.3.2 Genome reduction in intracellular pathogens, endosymbionts and marine α-proteobacteria

Mitochondria are strongly proposed to be descendents of bacteria, the α-proteobacteria in particular. The mitochondrial genome is highly reduced with a large number of genes required for its function derived from the host's nuclear genome. This loss of genetic information is likely to have happened as a result

---

44 Pearson M. M., Sebaihia M., Churcher C., Quail M. A., Seshasayee A. S., Luscombe N. M., Abdellah Z., Arrosmith C., Atkin B., Chillingworth T., Hauser H., Jagels K., Moule S., Mungall K., Norbertczak H., Rabbinowitsch E., Walker D., Whithead S., Thomson N. R., Rather P. N., Parkhill J. and Mobley H. L. 2008. 'Complete genome sequence of uropathogenic Proteus mirabilis, a master of both adherence and motility.' *Journal of Bacteriology* 190: 4027–4037.

45 Rasko D. A., Rosovitz M. J., Myers G. S., Mongodin E. F., Fricke W. F., Gajer P., Crabtree J., Sebaihia M., Thomson N. R., Chaudhuri R., Henderson I. R., Sperandio V. and Ravel J. 2008. 'The pangenome structure of *Escherichia coli*: Comparative genomic analysis of E. coli commensal and pathogenic isolates.' *Journal of Bacteriology* 190: 6881–6893.

46 Zakour, Guinane and Fitzgerald. 2008. 'Pathogenomics of the staphylococci: Insights into niche adaptation and the emergence of new virulent strains.' *FEMS Microbiology Letters* 289: 1–12.

of a population bottleneck in which deleterious, but non-lethal gene loss is not disfavoured because of a lack of selection.

Several bacterial pathogens of humans have adopted an intracellular lifestyle. An iconic example, at least from the perspective of genomic studies, is *Rickettsia prowazekii*, the agent of epidemic, louse-borne typhus in humans. Historically, this infectious disease killed several million people in the aftermath of the two World Wars. The infectious agent is an obligate intracellular parasite and appears to be the closest extant relative of the mitochondrion, on the basis of phylogenetic analysis of selected genes.

Siv Andersson and co-workers sequenced the ~1.1 Mb genome of *R. prowazekii*.[47] In contrast to other bacterial genomes, including those from the *E. coli* complex, a large proportion–24% (compared to ~10% for other bacterial genomes)–of the genome was non-coding (Fig. 2.8 a). Only a small portion

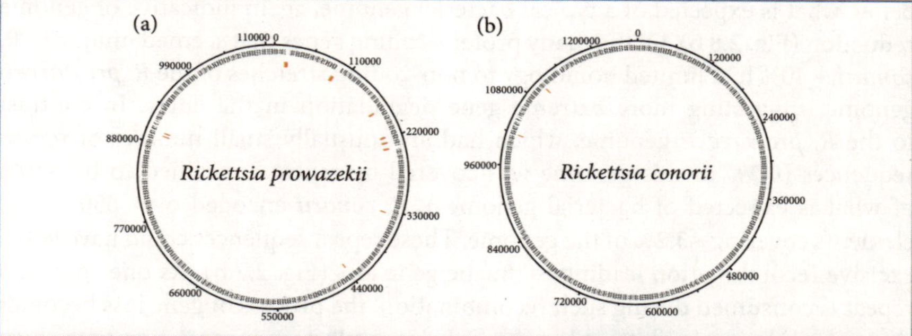

Figure 2.8 *Gene content of Rickettsia.* Genome tracks showing the gene content of the genomes of (a) *R. prowazekii* and (b) *R. conorii*. The grey track marks active genes, whereas the red track marks pseudogenes. Though there are only a few identifiable pseudogenes in *R. prowazekii*, even these are not present in *R. conorii*. The gene density of *R. conorii* is greater than that of *R. prowazekii*, though this may not be apparent from this figure. These were drawn using the DNA Plotter module packaged with the Artemis genome analysis software (https://www.sanger.ac.uk/resources/software/artemis/).

(~1%) could be attributed to pseudogenes, the rest may be gene relics degraded beyond recognition. Analysis of gene functions predicted in this genome showed a near-complete loss of amino acid biosynthetic enzymes, with the few remaining cases expected to feed into other essential cellular processes such as cell envelope

47 Andersson S. G., Zomorodipour A., Andersson J. O., Sicheritz-Pontén T., Alsmark U. C., Podowski R. M., Näslund A. K., Eriksson A. S., Winkler H. H. and Kurland C. G. 1998. 'The genome sequence of Rickettsia prowazekii and the origin of mitochondria.' *Nature*. 396: 133–43.

and cofactor metabolism. Thus this organism might import amino acids from its host. A similar import of nucleoside monophosphates was also implicated by such gene function analysis. Genes for anaerobic glycolysis were absent. However, as with mitochondria, the *R. prowazekii* genome encoded the pyruvate dehydrogenase complex and the full complement of the TCA cycle and the ATP synthase genes, suggesting import of pyruvate from the host cytoplasm. Further, many nuclear genes that contribute to mitochondrial function were encoded by the *R. prowazekii* genome suggesting that the mitochondrion is a more extreme example of genome reduction and has a more obligate relationship with the host than *R. prowazekii*.

What are the mechanisms of genome reduction? The genome of *Rickettsia conorii*, the causative agent of Mediterranean spotted fever, provided some clues.[48] The 1.3 Mb genome of *R. conorii* strongly conserved gene order with *R. prowazekii*. However, it had a higher coding potential than *R. prowazekii*–although it was far below what is expected of a typical bacterial genome, again indicative of genome reduction (Fig. 2.8 b). Of the many protein-coding genes that seemed unique to *R. conorii*, ~40% had limited homology to non-coding stretches of the *R. prowazekii* genome, suggesting more extreme gene degradation in the latter. In contrast to the *R. prowazekii* genome, which had an unusually small number of repeat sequences (0.2% of this genome was covered by repeats, specified to be ~10% of what is expected of bacterial genomes), *R. conorii* encoded over 650 repeat elements covering ~3.2% of the genome. These repeat sequences could have led to excisive recombination leading to drastic gene loss (Fig. 2.7 b). As one copy of a repeat is consumed during such recombination, the process of gene loss becomes slower and slower until it can happen only by subtle pseudogenisation with short intragenic deletions and mutation.[49] The reduction of *Rickettsial* genomes might have incorporated the whole gamut of these processes.

A second example of gene decay was seen in the genome of *Mycobacterium leprae*,[50] which causes leprosy, one of the oldest known human diseases.

---

[48] Ogata H., Audic S., Renesto-Audiffren P., Fournier P. E., Barbe V., Samson D., Roux V., Cossart P., Weissenbach J., Claverie J. M. and Raoult D. 'Mechanisms of evolution in Rickettsia conorii and R. prowazekii.' *Science* 293: 2093–2098.

[49] Andersson. 2004. 'Obligate intracellular pathogens.' *Microbial Genomics*, edited by Fraser, Read and Nelson. Humana Press, USA.

[50] Cole S. T., Eiglmeier K., Parkhill J., James K. D., Thomson N. R., Wheeler P. R., Honoré N., Garnier T., Churcher C., Harris D., Mungall K., Basham D., Brown D., Chillingworth T., Connor R., Davies R. M., Devlin K., Duthoy S., Feltwell T., Fraser A., Hamlin N., Holroyd S., Hornsby T., Jagels K., Lacroix C., Maclean J., Moule S., Murphy L., Oliver K., Quail M. A., Rajandream M. A., Rutherford K. M., Rutter S., Seeger K., Simon S., Simmonds M., Skelton J., Squares R., Squares S., Stevens K., Taylor K., Whitehead S., Woodward J. R. and Barrell B. G. 'Massive gene decay in the leprosy bacillis.' *Nature* 409: 1007–1011.

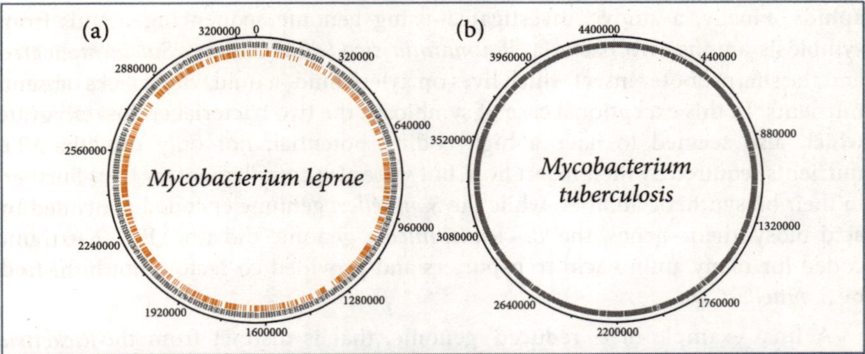

Figure 2.9 *Gene content of Mycobacteria.* Genome tracks showing the gene content of the genomes of (a) *M. leprae* and (b) *M. tuberculosis.* The grey track marks active genes, whereas the red track marks pseudogenes (none visible in *M. tuberculosis*). This figure highlights the large number of identifiable pseudogenes in *M. leprae,* indicative of a genome currently undergoing gene decay. These were drawn using the DNA Plotter module packaged with the Artemis genome analysis software.

Gene decay in this ~3.3 Mb genome is even more dramatic than in *R. prowazekii* (Fig. 2.9). While a bacterial genome of ~3.3 Mb typically encodes about 3,000 protein-coding genes, the genome of *M. leprae* coded for only ~1,600. In fact, a little over 50% of the genome was non-coding, of which over half could be attributed to pseudogenes. In contrast to the *Rickettsiae, M. leprae* encoded many amino acid biosynthetic enzymes, probably reflecting the non-availability of this nutrient in the host environment of the phagosome. Comparison with the genome of the related *Mycobacterium tuberculosis* first revealed major rearrangements reflected by changes in gene order. This comparison also showed a process of genome streamlining, where multi-gene families in the *M. tuberculosis* genome were reduced to a single copy in *M. leprae,* with some of the other copies identifiable as pseudogenes.

The authors of the report describing the *M. leprae* genome comment that "... it is striking that elimination of pseudogenes by deletion lags far behind inactivation ..." in both *Rickettsiae* and *M. leprae.* This is exemplified in contrast by the genome of *Buchnera sp.,*[51] a close but genome-reduced relative of *E. coli,* and an obligate endosymbiont of aphids, wherein 88% of the small 0.6 Mb genome represented protein-coding genes. This genome also demonstrated symbiosis, wherein *Buchnera* appeared to provide nutrients to its host, by encoding a full complement of genes for the biosynthesis of those amino acids essential to

[51] Shigenobu S., Watanabe H., Hattori M., Sakaki Y. and Ishikawa H. 2000. 'Genome sequence of the endocellular bacterial symbiont of aphids buchnera sp. APS.' *Nature.* 407, 6800: 81–6.

aphids. Finally, a study[52] investigated–using genome sequencing–a three-way symbiosis among two bacteria–*Baumannia cicadellinicola* and *Sulcia muelleri*–and the sharpshooter insect which lives on xylem fluid–a fluid which lacks organic nutrients. In this exceptional case of symbiosis, the two bacterial endosymbionts, which also seemed to have a high coding potential, not only provided the nutrients required by their insect host, but were also complementary to each other in their biosynthetic abilities; while the *S. muelleri* genome encoded many amino acid biosynthetic genes, the *B. cicadellinicola* genome did not (Fig. 2.10), but coded for many amino acid transporters and provided co-factors not produced by *S. muelleri*.

A final example of a 'reduced' genome, that is distinct from those of the endosymbionts is that of the marine α-proteobacterium, *Pelagibacter ubique*. The ~1.3 Mb genome was sequenced in 2005[53] and shown to be the smallest among all free-living bacteria. The lack of selection due to population bottlenecks, which has been proposed as the mechanism for genome reduction in endosymbionts is not applicable here. This is because *P. ubique* belongs to a class of organisms that are among the most successful in the planet, accounting for up to 25% of the total bacterial content of the oceans. Living in a nutrient-poor environment, this organism appears to code for most metabolic capabilities seen in α-proteobacteria, such as amino acid and fatty acid biosynthesis. Consistent with its notable ability to consume a large fraction of the dissolved organic carbon content of their marine environment, the genome was annotated to be rich in efficient broad specificity ABC transporters, with specific transporters–again of the ABC class–for certain nitrogenous compounds and osmolytes. Despite being well-endowed in these resources, the genome has undergone 'streamlining', where selection against the metabolic burden of replicating non-productive DNA has led to genome reduction. This is reflected in a remarkable paucity of duplicated genes, a lack of repetitive DNA and the absence of pseudogenes or evidence of recent horizontal gene acquisition including phages. This is also demonstrated by the exceptionally high coding density, with a median intergenic spacer size of only 3 bp, compared to ~85 bp for *E. coli*. Though the high A/T content of other reduced genomes have been attributed to specific mutational pressures, a similar base composition in *P. ubique* has been hypothesised to be an optimisation designed to reduce the metabolic requirement of limiting nitrogen.

---

52 Wu D., Daugherty S. C., Van Aken S. E., Pai G. H., Watkins K. L., Khouri H., Tallon L. J., Zaborsky J. M., Dunbar H. E., Tran P. L., Moran N. A. and Eisen J. A. 2006. 'Metabolic complementarity and genomics of the dual bacterial symbiosis of sharpshooters.' *PLoS Biology* 4: e188.

53 Giovannoni S. J., Tripp H. J., Givan S., Podar M., Vergin K. L., Baptista D., Bibbs L., Eads J., Richardson T. H., Noordewier M., Rappé M. S., Short J. M., Carrington J. C. and Mathur E. J. 2005. 'Genome streamlining in a cosmopolitan oceanic bacterium.' *Science*. 309: 1242–1245.

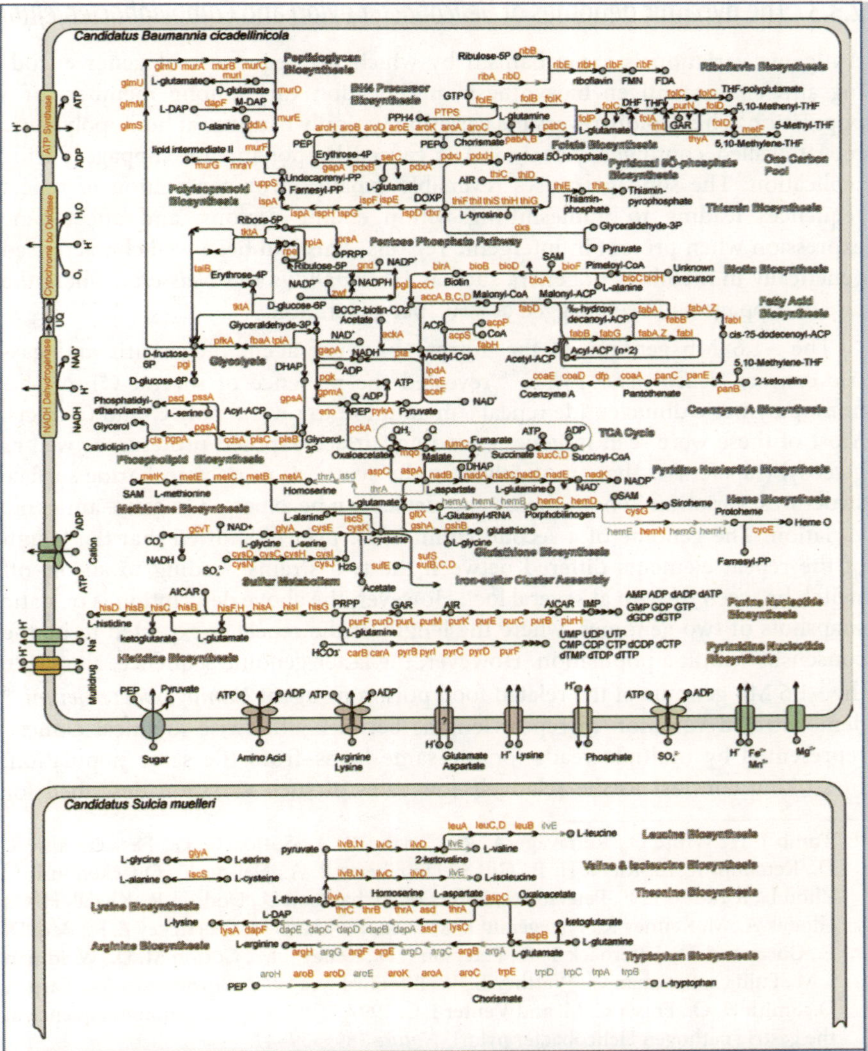

Figure 2.10 *Metabolic potential of two co-operating bacteria.* The genome of *Baumannia cicadellinicola* does not encode pathways for the biosynthesis of many amino acids, which are instead provided by its symbiotic partner *Sulcia muelleri*. *B. cicadellinicola* encodes many amino acid importers. This figure is reproduced under the Creative Commons Attribution licence from Wu D., Daugherty S. C., Van Aken S. E., Pai G. H., Watkins K. L., et al. 2006. 'Metabolic Complementarity and Genomics of the Dual Bacterial Symbiosis of Sharpshooters.' *PLoS Biol* 4, 6: e188.:10.1371/journal.pbio.0040188. © Wu et al. 2006.

2.3.3 The dynamic genomes of *Helicobacter pylori* and *Campylobacter jejuni*

Antigenic variation is a mechanism by which the sequences of genes encoding antigens or antigen biosynthesis/modification vary among members of a population. This rapidly reversible variation typically happens at homopolymeric repeats called contingency loci, and is caused by polymerase slippage during replication. The slippage causes reversible expansion or contraction of repeat sequences leading to frameshifting within coding regions, and affect gene expression when present in intergenic regions. This had been well characterised genetically in members of genera such as *Haemophilus* and *Neisseria*, where the rate of slippage had been pegged at ~10^{-3} per generation.

The ~1.6 Mb genome of the ε-proteobacterial agent of gastritis and gastric cancer, *Helicobacter pylori*,[54] revealed the presence of several CT or AG homopolymeric dinucleotide repeats, in addition to several poly C-and G-tracts. Most of these were seen in genes encoding surface-exposed proteins, as well as glycosyl transferases that are involved in the synthesis of polysaccharide surface structures; variation in both these gene classes can be expected to cause antigenic variation. The genome of a second strain of *H. pylori*[55] showed that the length of the repeat elements differed between the two strains, leading to an on–off switch between the two at several loci. However, the above description is of static snapshots of two genomes, where the length of the repeat is expected to be the consensus within a population. However, the latter genome sequence, alongside the ~1.6 Mb genome of the related food poisoning agent *Campylobacter jejuni*,[56] demonstrated variation in repeat lengths between otherwise identical clones–represented by multiple reads for the same locus–from the same population. In striking contrast to the relatively low rates of such variation described for

[54] Tomb J. F., White O., Kerlavage A. R., Clayton R. A., Sutton G. G., Fleischmann R. D., Ketchum K. A., Klenk H. P., Gill S., Dougherty B. A., Nelson K., Quackenbush J., Zhou L., Kirkness E. F., Peterson S., Loftus B., Richardson D., Dodson R., Khalak H. G., Glodek A., McKenney K., Fitzegerald L. M., Lee N., Adams M. D., Hickey E. K., Berg D. E., Gocayne J. D., Utterback T. R., Peterson J. D., Kelley J. M., Cotton M. D., Weidman J. M., Fujii C., Bowman C., Watthey L., Wallin E., Hayes W. S., Borodovsky M., Karp P. D., Smith H. O., Fraser C. M. and Venter J. C. 1997. 'The complete genome sequence of the gastric pathogen Helicobacter pylori.' *Nature* 389: 539–47.

[55] Alm R. A., Ling L. S., Moir D. T., King B. L., Brown E. D., Doig P. C., Smith D. R., Noonan B., Guild B. C., deJonge B. L., Carmel G., Tummino P. J., Caruso A., Uria-Nickelsen M., Mills D. M., Ives C., Gibson R., Merberg D., Mills S. D., Jiang Q., Taylor D. E., Vovis G. F. and Trust T. J. 1999. 'Genomic-sequence comparison of two unrelated isolates of the human gastric pathogen Helicobacter pylori.' *Nature* 397: 176–80.

[56] Parkhill J., Wren B. W., Mungall K., Ketley J. M., Churcher C., Basham D., Chillingworth T., Davies R. M., Feltwell T., Holroyd S., Jagels K., Karlyshev A. V., Moule S., Pallen M. J., Penn C. W., Quail M. A., Rajandream M. A., Rutherford K. M., van Vliet A. H., Whitehead S. and Barrell B. G. 2000. 'The genome sequence of the food-borne pathogen Campylobacter jejuni reveals hypervariable sequences.' *Nature* 403: 665–68.

Haemophilus and *Neisseriae* (see above), multiple variants at the same contingency locus were equally populated among the shotgun reads, suggesting variation that is several orders of magnitude higher than anticipated. This high level of variation was attributed to the notable absence of several proteins involved in DNA repair, many of which are encoded by the genome of *H. influenzae*, also known to undergo antigenic variation at contingency loci, albeit at lower frequencies.

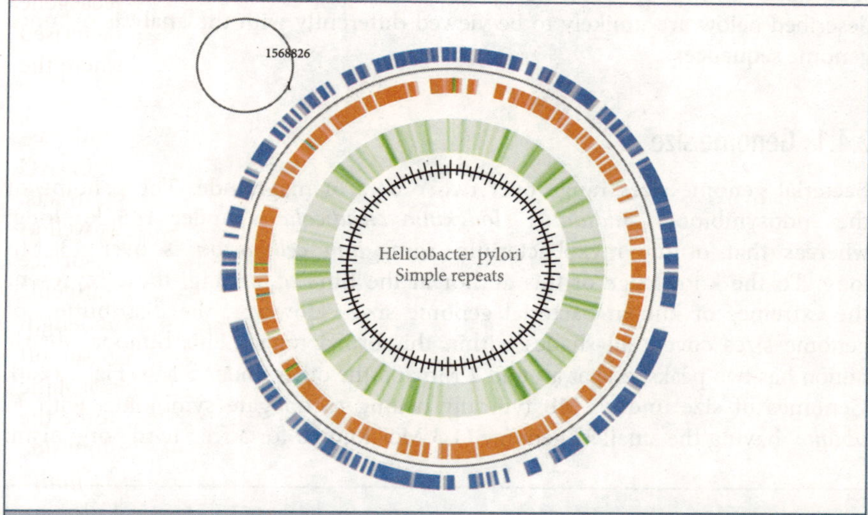

Figure 2.11 *Clusters of simple repeats in the genome of H. pylori genome.* The genome track in this figure shows the presence of several clusters of simple sequence repeats (shown by green patches in the inner band) in the *H. pylori* genome. This was drawn using the GeneWiz browser.

2.4 Some lessons learnt from studying 2,000 bacterial genomes

While lessons learnt from individual genomes have been instructive, it is clear that most insights–within the ambit of a genome sequencing and annotation project–into the biology of the organism has largely come from making comparisons with other genomes, be it the role of horizontal gene transfer in *E. coli* virulence, or distinct priciples of genome reduction underlying the evolution of *Rickettsiae* and *P. ubique*, or the dynamic nature of the genomes of *H. pylori* and *C. jejuni*. The above-described limited comparisons provided significant insights into the structure and evolution of the genomes in question. However, general trends underpinning the organisation and evolution of bacterial genomes have been derived largely through meta-analysis of many genomes from across diverse taxonomic groups. These general principles have been reviewed extensively by Koonin and Wolf.[57] Here, we present a selection of the most striking observations. There are over 3,000 complete bacterial genomes in public databases. Though there are strong biases in the selection of organisms whose genomes have been sequenced–for example, medically-relevant bacteria–many of the properties described below are unlikely to be viewed differently with the analysis of more genome sequences.

2.4.1 Genome size

Bacterial genome sizes range over two orders of magnitude. The genome of the endosymbiont *Candidatus Hodgkinia cicadicola* is under 145 kb long, whereas that of the myxobacterium *Sorangium cellulosum* is over 13 Mb long. To the knowledge of this author at the time of writing, these represent the extremes of known bacterial genome sizes. However, the distribution of genome sizes covers all shades within this broad range. This bimodal distribution has two peaks, one at around 2 Mb and the other under 5 Mb (Fig. 2.12 a). Genomes of size under 1Mb typically belong to obligate symbionts, with *P. ubique* having the smallest genome (1.3 Mb) known for a free-living organism.

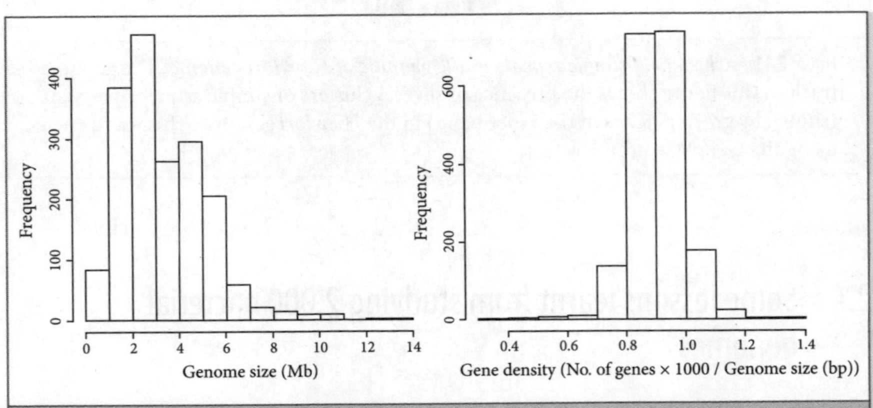

Figure 2.12 *Genome size and gene density in bacteria.* The distributions of (a) genome sizes and (b) gene density – across ~3,000 fully-sequenced bacterial genomes – are shown here as histograms.

[57] Koonin and Wolf. 2008. 'Genomics of bacteria and archaea: The emerging dynamic view of the prokaryotic world.' *Nucleic Acids Research* 36: 6688–6719.

As discussed in the case studies above, several evolutionary forces could have influenced genome sizes:

1 *Innovation - horizontal gene transfer and gene duplication*: As noted for the *E. coli* complex, this can lead to drastic variations in genome sizes even within a species. Large-scale horizontal gene transfer, except under extreme conditions where it leads to subsequent massive gene loss, increases genome sizes. In addition to gene transfer across bacteria, these also include elements such as bacteriophages and plasmids which add to an increase in genome size when integrated into the main chromosome. Besides this, gene duplication also leads to the formation of multi-gene families, where members might share similar functions, or be diverged substantially to have discovered new functions.

2 *Genome reduction - population bottlenecks and selection-driven gene loss*: As described earlier, genome reduction can happen under a population bottleneck where a small population–having lost a set of non-lethal, but deleterious genes–survives because of the absence of selection against it; this is likely to have happened in endosymbionts such as the mitochondria and the related intracellular pathogen *Rickettsia*. On the other hand, certain genomes such as *P. ubique* have possibly undergone genome reduction as a part of streamlining their genome in the face of selection from nutrient limitation.

Various gene functions that could impose ceilings on genome sizes are described later.

2.4.2 Coding density

In contrast to many eukaryotes, bacteria (and archaea) have high protein-coding densities, with a typical number of 0.9 genes per kilobase of genomic sequence (Fig. 2.12 b). Intergenic regions are small, with the occasional long ones containing specialised non-coding elements such as the CRISPR repeats which are involved in limiting horizontal gene transfer, and pseudogenes or relics as described above in organisms such as *M. leprae* and *R. prowazekii*. Some genomes have extremely short intergenic regions, sometimes averaging <10 bp. As described above, one such example is the streamlined genome of *P. ubique*. Yet another example is *C. jejuni* which, similar to *P. ubique*, has a small genome with little evidence of fixed horizontally-acquired islands. To the knowledge of this author, genome streamlining in *C. jejuni* has not been formally stated, nor has it been tested whether extremely short intergenic regions are predictive of genome streamlining, which may well be true.

Coding densities are generally higher in the leading strand than in the lagging strand. Highly expressed genes are generally encoded such that transcription and replication at these loci progress in the same direction. These could be because of a need to minimise collisions between the replicating and the transcribing polymerases.

2.4.3 Gene order conservation

The order in which genes are encoded is generally not conserved on a genomic scale (Fig. 2.13). Comparisons of even closely related genomes show that co-linearity in gene order is broken at several points. Between moderately divergent bacteria, there are only short stretches of collinearity with the possibility of recombination in replication forks leading to large-scale inversions. At larger evolutionary distances, there is hardly any sign of gene order conservation. However, it is to be noted that many operons, which are groups of genes encoded next to each other and typically form a single transcription unit, are conserved in the sense that they encode the same genes across several genomes. Most operons contain only 2–4 genes, however, the above is also true of the much larger 'uber-operons', such as the one that encodes many ribosomal proteins. Though the gene content of an operon may be conserved between two genomes, as a result–for example–of a selective pressure for co-expression of genes encoded therein, the order in which the genes that constitute the operon are encoded may not be maintained.

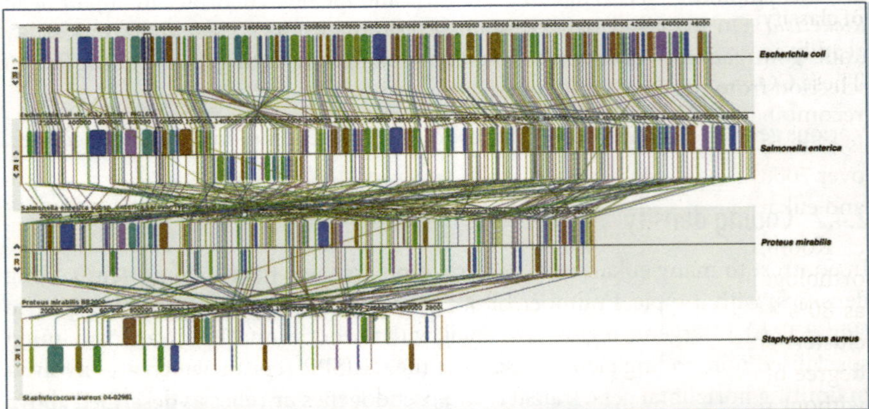

Figure 2.13 *Gene order conservation.* Genome-wide sequence alignments of *E. coli K12* (top) with a member of the related species *Salmonella enterica*, a more distant enterobacterial relative *Proteus mirabilis*, and the very distant *Staphylococcus aureus* (bottom) are shown. Each genome is drawn as a horizontal line with genes marked; the length of the horizontal line is proportional to the size of the genome. Positive alignments across genomes are represented by thin lines extending from one genome to another. Though the number of genes that is conserved between *E. coli* and S. *aureus* is obviously less than that between *E. coli* and *S. enterica*, it is to be noted that even for the conserved genes synteny is poorly maintained in the former pair when compared to the latter. This figure was drawn using the Mauve software (http://gel.ahabs.wisc.edu/mauve/).

2.4.4 Comparative genomics of gene functions: Systematic annotation

Comparative genomics has allowed the classification of genes into what are called *functional categories*. For example, a component of the ribosome is classified under *translation*, and a gene involved in the biosynthesis of lysine under *amino acid biosynthesis*. These categories can be further refined to call the gene involved in amino acid biosynthesis as being involved in *lysine metabolism* and even further as one with a *diaminopimelate decarboxylase* activity. At this finest level, it is equivalent to the most visible annotation one sees for a gene in a database such as Genbank. Such annotations and classifications, arrived at in a systematic and consistent manner, are available in databases such as GO,[58] COG[59] and EggNOG,[60] each database using a distinct annotation procedure.

The COG database is probably the most popular functional annotation database for prokaryotes. Each COG, which stands for a cluster of ortholgous group, is a set of three or more proteins which are deemed to be orthologous to each other. It includes both fast-and slow-evolving families of proteins. A step-wise description of classifying proteins into COGs is available elsewhere.[61] The COG database that is readily accessible online covers ~5,000 COGs covering ~66 prokaryotic genomes. These COGs are merged into ~20 broad functional categories such as 'replication, recombination and repair' and 'nucleotide transport and metabolism'. EggNOG is a more comprehensive version of the COG database, and currently includes over 700,000 orthologous groups from more than 1,100 species, both prokaryotes and eukaryotes.

Koonin and Wolf,[62] in an analysis of the EggNOG database comprising ~30,000 orthologous groups from ~350 bacterial and archaeal genomes, state that as many as 80% of genes in any genome can be readily annotated with an orthologous cluster. Thus a large proportion of genes in any bacterial genome shows a certain degree of evolutionary conservation, thus allowing us to assign protein function without requiring detailed experiments. However, it is to be noted that as many as 80% of orthologous groups are found in only a few bacterial genomes; most of the remaining groups are only moderately common with a very small number

---

58 http://www.geneontology.org/, accessed on 9th July, 2014.

59 http://www.ncbi.nlm.nih.gov/COG/, accessed on 9th July, 2014.

60 http://eggnog.embl.de/version_3.0/index.html, accessed on 9th July, 2014.

61 Koonin. 2002. 'The clusters of orthologous groups (COGs) database: Phylogenetic classification of proteins from complete genomes.' *The NCBI Handbook*, edited by McIntyre, and Ostell. National Centre for Biotechnology Information.

62 Koonin and Wolf. 2008. 'Genomics of bacteria and archaea: The emerging dynamic view of the prokaryotic world.' *Nucleic Acids Research* 36: 6688–6719.

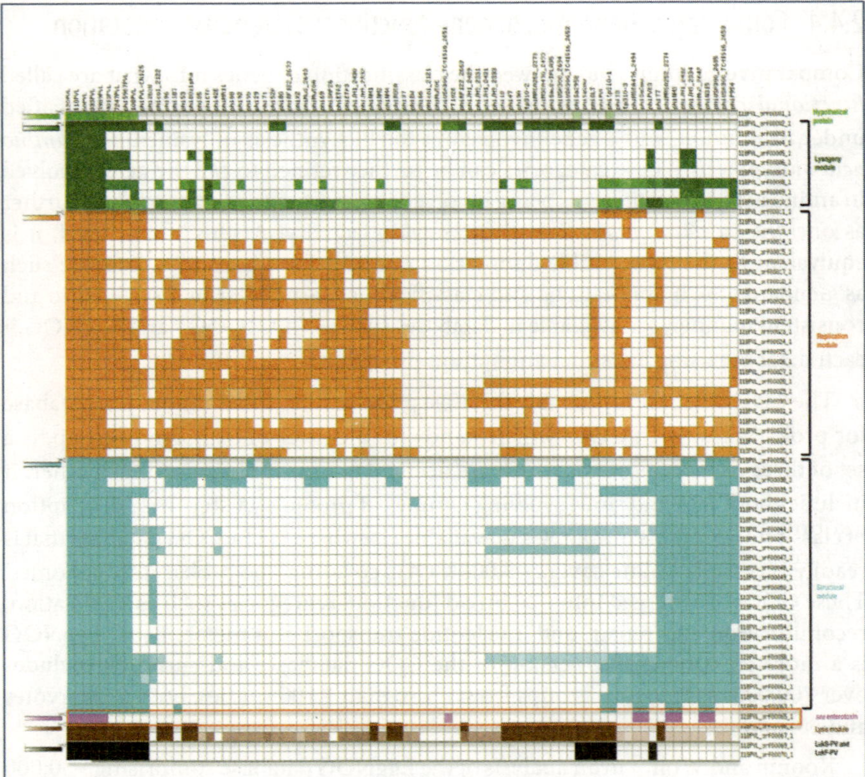

Figure 2.14 *Genes conserved across S. aureus prophages.* This figure is an example of a heat map showing the presence or absence of a gene across genomes. Here the genome is the toy example of a class of prophages present in *S. aureus* genomes. ORFs (represented on the vertical axis) from a prophage present in an Indian isolate of *S. aureus* are compared to their corresponding orthologs in other *S. aureus* prophages (represented on the horizontal axis). Genes are color coded as per the functional class they belong to (marked in the figure). Intensity of the color indicates the percent sequence similarity between any two. The *sea* enterotoxin gene, by being in two distantly related types of prophages but not in others, suggests possible recombination between the two types of phages; this gene has been highlighted by a red box. This figure is reproduced under the Creative Commons Attribution license from Prabhakara S., Khedkar S., Shambat S. M., Srinivasan R., Basu A., et al. 2013. 'Genome Sequencing Unveils a Novel Sea Enterotoxin-Carrying PVL Phage in Staphylococcus aureus ST772 from India.' *PLoS ONE* 8, 3: e60013. doi:10.1371/journal.pone.0060013'. © Prabhakara et al. 2013.

(~0.2%) being present in most bacteria. In other words, most gene sequences are not conserved across bacteria, suggesting that the minimal genome of bacterial life is very small.

The systematic classification of genes into orthologous sets allows us to use genome sequences to quickly group organisms on the basis of the gene functions they encode. This is done by first representing a genome as a vector, where each entry represents the presence or absence of a particular orthologous group in that genome. The matrix thus obtained is clustered, wherein genomes with similar ortholog occurrence vectors are placed next to one another (Fig. 2.14). Such clustering places taxonomically related organisms close to each other with deviations caused by forces such as lineage-specific gene decay and large-scale gene acquisitions by horizontal transfer.

2.4.5 Comparative genomics of gene functions: Scaling laws

Besides the various obvious benefits of knowing the function(s) of a protein–even at a broad level–such classification have also permitted innovative thoughts on the forces that constrain genome sizes. This follows from the question: "What is the relative occurrence of each broad-level functional category in a genome, and how does this occurrence scale with genome size?".

van Nimwegen performed such a study[63] for broad-level functional categories defined in the GO database and showed that the relationship between genome size and the number of genes from any functional category is a power law of the form $n_c = \lambda g^r$, where n_c is the number of genes from a functional category, g the total number of genes encoded by the genome, and r the exponent with λ being a constant. The exponent r has been interpreted as the ratio of the effective duplication rate for genes from the appropriate functional category and that for the genomic average. For core processes such as DNA replication and translation, r is well under 1, suggesting that for such categories there is only a small–if any–increase in gene numbers with genome size. This is similar to the observation made while annotating the genome of *E. coli* K12. Metabolic systems grow linearly with genome size, such that the proportion of genes for metabolism is nearly constant. Finally, regulatory proteins have an $r \sim 2$ showing that the number of regulators required, on average, for a gene increases with genome size (Fig. 2.15). Models for the emergence of the quadratic scaling of regulatory proteins exist,[64] but these are not discussed here.

---

63 van Nimwegen. 2003. 'Scaling laws in the functional content of genomes.' *Trends in Genetics* 19: 479–84.

64 For example, Maslov S., Krishna S., Pang T. Y. and Sneppen K. 2009. 'Toolbox model of evolution of prokaryotic metabolic networks and their regulation.' *Proceedings of the National Academy of Sciences USA* 106: 9743–9748.

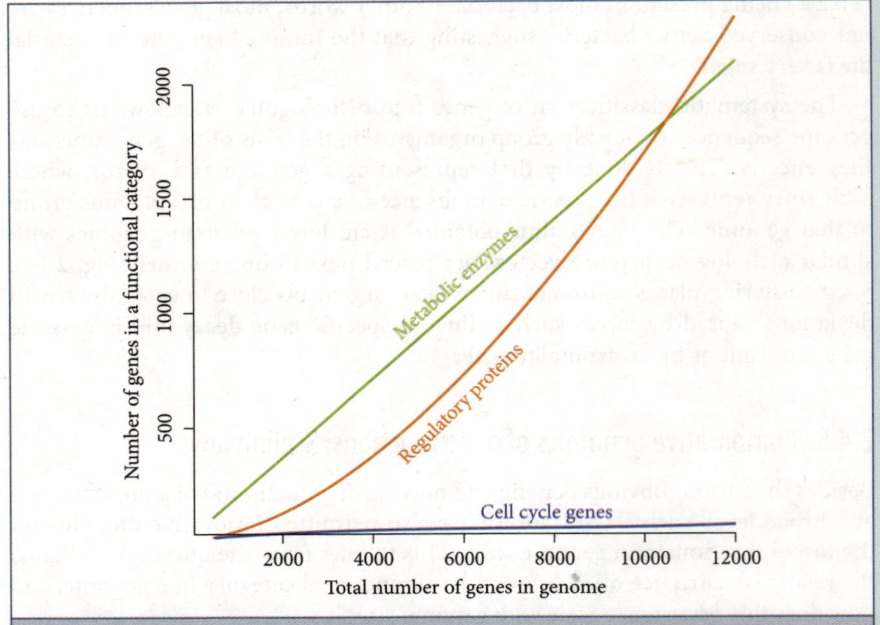

Figure 2.15 *Scaling laws.* Scaling of three classes of genes with total number of genes in a genome. The curves show only the fits, with the equations for metabolic and regulatory proteins taken from Ranea et al. (2005), and cell cycle proteins from van Nimwegen (2003).

Using an analogy to a factory, Ranea and colleagues[65] proposed that the super-linear growth of regulatory proteins, compared to the linear scaling of metabolic genes with genome size, could impose a ceiling on genome size itself. These authors equate metabolic enzymes to a profitable output and regulatory proteins to the logistic cost associated with producing that output. Where the cost exceeds benefit lies the optimal upper limit for genome size. The scaling factors described for metabolic enzymes and regulatory proteins showed that the genome size ceiling is at ~10,500 genes, which as an approximation agrees reasonably well with the known distribution of prokaryotic genome sizes. Adrian Bird[66] had previously proposed that novel regulatory mechanisms, will be required for effective regulation of genes in large genomes, which is particularly true of complex eukaryotes.

---

65 Ranea J. A., Grant A., Thornton J. M. and Orengo C. A. 2005. 'Microeconomic principles explain an optimal genome size in bacteria.' *Trends in Genetics* 21: 21–25.

66 Bird. 1995. 'Gene number, noise reduction and biological complexity.' *Trends in Genetics* 11: 94–100.

Summary

- ✓ Bacterial genomes are sequenced and assembled using a whole genome shotgun sequencing strategy.
- ✓ Bacterial genomes can be annotated with protein-coding and RNA genes, and to a lesser extent with regulatory elements, using computational methods.
- ✓ Bacterial genomes expand through duplication and horizontal gene transfer and reduce by decay/deletion under a population bottleneck or under environments selecting for streamlining.
- ✓ Though many bacterial genes show some degree of evolutionary conservation, gene order is not conserved due to rampant recombination.
- ✓ The prevalence of different gene functions in a genome scale differently with genome size, and this can be used to predict an upper limit on prokaryotic genome sizes.

3

Studying Bacterial Genome Variation with Microarrays

3.1 Introduction

The sequencing of even a few members of a bacterial species has underlined the remarkable genetic diversity that underpins these organisms. These have suggested that in many taxa, there could be innumerable genetically unique members and that their genetic characterisation is best initiated by obtaining information on those genomic attributes that are unique to each member. However, genome sequencing using the Sanger methodology–described in the previous chapter–is expensive and time-consuming, and therefore not a practical option when it comes to sequencing 'every' genetically unique isolate of some relevance. However, various alternatives have been remarkably successful, and these have led to the field of 'phylogenomics' or 'genomic epidemiology'.

3.2 DNA microarrays: The concept

The first approach that permits 'phylogenomics' is the use of DNA microarrays to probe sequence variations between different isolates.

DNA microarrays are based on the concept of hybridisation of a nucleic acid to one of complementary sequence–a fundamental principle underlying many molecular biology methods for detecting and quantifying nucleic acids of a defined sequence. These microarrays are slides containing many nucleic acid probes, allowing thousands to millions of hybridisation experiments to be run in parallel. The technology can be used to detect the presence and absence of genes or more subtle polymorphisms in a genome (comparative genome hybridisation or **CGH** , the subject of this chapter), calculate relative expression levels of all genes encoded in a genome and compare these levels across several conditions and genetic backgrounds (gene expression microarray), and even measure, semi-quantitatively, the levels of binding of a DNA-binding protein to various parts of the genome (chromatin immuniprecipitation chip or ChIP-chip).

Irrespective of the application, there are certain general points to be noted while analysing and interpreting microarray data. A few of these–they are not exhaustive–are as follows:

1 *The general idea of a microarray:* In a DNA microarray experiment, fluorescently-labelled nucleic acid sample is hybridised against unlabelled, complementary probe sequences. The probes are pre-designed and pre-fixed on the slide with the labelled sample being added to the slide during the experiment. As the unhybridised excess sample is washed off, the fluorescence intensity obtained from each probe is proportional to the amount of sample hybridised to it. However, various issues including background noise and the difficulty of performing absolute quantification, necessitate careful data processing for meaningful biological information to be extracted. The essence of these methods is briefly described in subsequent sections of this chapter.

2 *What is probed by the microarray:* Many microarrays have probes covering only annotated genes. Many others, typically those interrogating samples from genomes that have not been sequenced, probe representative gene sequences from across multiple fully-sequenced genomes of related organisms. These arrays do not interrogate intergenic regions, or portions of the genome that have been erroneously not annotated as genes. A second class of microarrays, termed '*tiling microarrays*', cover the entire genome with oligonucleotide probes. These tiling arrays can and have been used to provide experimental annotation of genomes; this is discussed in a later chapter. Issues related to the design of microarrays are briefly mentioned in Sections 3.5.1 and 3.5.2, in the context of CGH experiments.

3 *One-or two-channel arrays:* Some microarrays, called *two-channel* arrays, allow the hybridisation of two samples, one from a reference and the other from a test sample, each labelled with a differently coloured fluorescent probe (R for red, and G for green). Others, especially those from Affymetrix, permit the hybridisation of only a single sample and therefore, detect signals from only one *channel*. Hence, experiments comparing nucleic acid abundance between two samples can be done using a single microarray slide in a two-channel format, but require one for each sample when a one-channel platform is adopted. It also goes without saying that there will be differences in the manner in which data obtained from the two different types of arrays are processed.

3.3 DNA microarrays: From fluorescence intensities to information

A microarray experiment ultimately produces fluorescence intensities which are proportional to the amount of nucleic acid hybridised to the corresponding probe. Each hybridised probe may produce fluorescence in one or two channels depending on the type of experiment. The raw data thus produced have to be computationally processed before it can be used as a meaningful representation of the required biological information. A typical data processing pipeline is illustrated in Fig. 3.1. More detailed description of these methods are available in

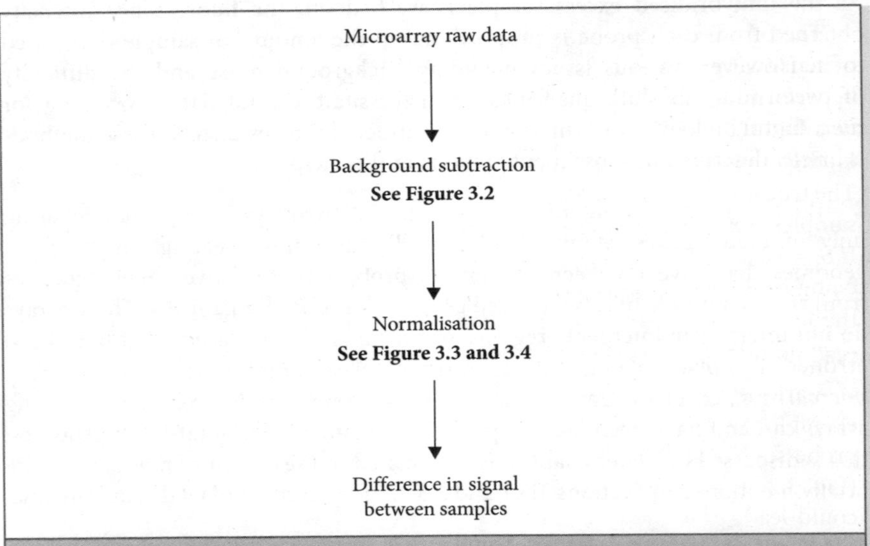

Figure 3.1 *Steps involved in a typical microarray data processing pipeline.* Irrespective of the application to which a microarray is used for, certain steps need to be followed. First, background fluorescence should be removed (see Fig. 3.2), which helps in enhancing the sensitivity of the array to measuring differences in signal between samples. Data across channels and/or arrays should be normalized to the same *scale*, to enable direct comparisons (see Figs 3.3 and 3.4); many state-of-the-art normalization methods assume that a majority of probes do not display a significant change in signal between the samples compared. These normalized data may be used as such, but more typically signals obtained from each probe are compared across samples to statistically test whether the samples compared differ significantly from each other in terms of the signal obtained from a probe (a stock term that is used for this process is *differential expression*).

other books.[1] Readers interested in performing these analyses at an advanced level are directed to the primary literature and to open source software packages such as *Bioconductor*,[2] implemented in the free statistics suite *R*.[3] Guided tutorials comprising various aspects of what is described below, and more, are available in various books[4] and in the documentation for bioconductor packages.

3.3.1 Background correction

All fluorescence intensities from a microarray experiment include significant background. This is generally attributed to fluorescence due to non-specific hybridisation of the sample to non-cognate probes, and the fluorescence intrinsic to the glass slide itself.

The problem with having background noise is that it causes a significant loss of sensitivity when computing differences in the signal from a particular probe between two different samples. For example, if the true signal from a probe for two different samples are S_1 and S_2, and the respective backgrounds are b_1 and b_2, then the measured fluorescence intensities are $S_1 + b_1$ and $S_2 + b_2$, respectively. The true fold change in abundance of the hybridised nucleic acid between the two samples is S_1/S_2; however, with the background included, it is only $S_1 + b_1/S_2 + b_2$. If $S_1 = 10$ and $S_2 = 20$ and $b_1 = b_2 = 5$, then a true fold change of sample 2 ends up being measured as only 1.67. The higher the background fluorescence intensity, the lower the fold change measured, and therefore lower the sensitivity of the readout to differences in concentration of the hybridising nucleic acid.

Various methods are available for background subtraction (Fig. 3.2). For example, in two-colour arrays, fluorescence from a set of pixels surrounding each probe spot is used to calculate the channel-specific background for each probe (Fig. 3.2 a). However, erroneous inclusion of pixels containing the probe itself could lead to over-estimation of the background. Further, this method cannot account for non-specific hybridisation between non-cognate sequences, which can however be estimated from the fluorescence of control probes not expected to hybridise with any portion of the source genome. In Affymetrix arrays, every *perfect match* probe has a *mismatch probe*, with a one-base mismatch at the middle of the probe sequence. In theory, any fluorescence seen from these probes

---

1 (a) Stekel. 2003. *Microarray Bioinformatics*. Cambridge University Press: USA; (b) Causton, Quackenbush and Brazma. 2003. Microarray: *Gene Expression Data Analysis*. Blackwell publishing: UK.

2 http://www.bioconductor.org

3 http://www.r-project.org

4 (a) Gentleman, Carey, Huber, Irizarry and Dudoit. Eds. 2005. 'Bioinformatics and computational biology solutions using R and bioconductor.' *Springer*; (b) Hahne, Huber, Gentleman and Falcon. Eds. 2008. 'Bioconductor case studies.' *Springer*.

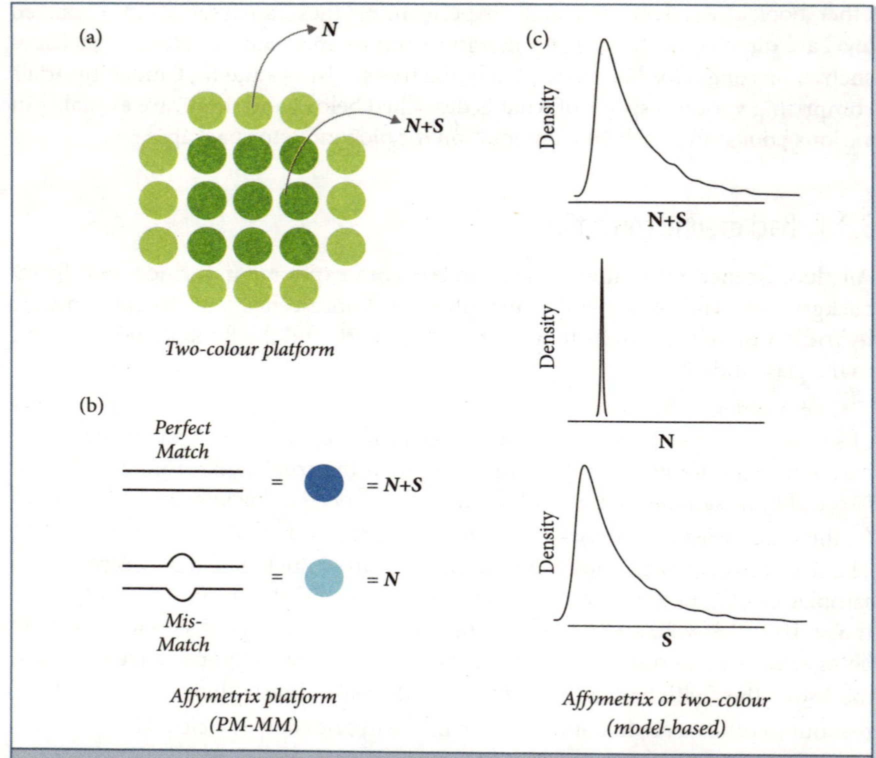

Figure 3.2 *Background subtraction for microarray data.* Schematic representations of various background subtraction methods for microarray data are shown. *N* represents the background and *S* the signal; the data obtained from the microarray is *N+S*. (a) Background subtraction for two-colour microarray data using intensities of pixels surrounding those containing the probe; (b) a simple method of background correction for Affymetrix data, wherein the signal from a *perfect-match* probe is subtracted from that from the corresponding mismatch probe, which has a single-base mismatch – relative to the *perfect-match* probe – in the middle of the probe sequence; (c) model-based background correction, wherein the data are represented as a sum of background, which is represented by a normal distribution, and the signal, approximated by an exponential distribution. The distributions shown were not derived from real data, but from a simulated example.

should correspond to the background and can be subtracted from the fluorescence obtained from the corresponding perfect match probe (Fig. 3.2 b). However, for several probes, the mismatch intensity ends up being higher than the corresponding perfect match intensity, producing meaningless negative signals. Though it can be reasonably argued that such probe pairs should be ignored, because the fluorescence signal is unreliable, or because the abundance of the hybridised nucleic acid is too low to be detected robustly, various methods (see below) are able to avoid negative signals while not discarding these probes.

Some state-of-the art methods, which are implemented in the open source Bioconductor suite of tools, completely ignore mismatch probe data from Affymetrix arrays. Instead, they fit a mixed distribution to the perfect match signals so that a normal distribution component corresponds to the background, with the signal being approximated by an exponential distribution (Fig. 3.2 c). Similar methods can be applied to two-channel array data as well. These methods are included in the *RMA* algorithm for Affymetrix data,[5] and *normexp* for two-channel data.[6]

3.3.2 Normalisation

In a two-channel microarray, the same array produces data from two lasers operating at different frequencies. This might cause systematic differences between the fluorescence intensities from the two channels. These differences are not biologically meaningful and have to be corrected. Several methods, together classfied under *within-array normalisation*, are available for this purpose. To describe these methods, let us define the background-subtracted signal from the reference green channel as G, and that from the red channel as R. Our objective is to measure the fold change between R and G, typically expressed on a $\log_2$ scale as $\log_2 R - \log_2 G$.[7]

In the first method, $\log_2 R$ is graphed against $\log_2 G$ as a scatter plot, with the former on the y-axis (Fig. 3.3 a). In the absence of any artifacts, one would expect the best fit regression line to have a slope of 1 and a y-intercept of 0. However, this is not the case, with both the slope and the intercept deviating from the ideal. Standard statistics can be used to adjust for this. First, the slope (m) and the y-intercept (c) of the regression line are calculated. Then $\log_2 R$ is scaled, so that it is now $\log_2 R' = \log_2 (R - c)/m$ with $\log_2 G$ left untouched. This gives the normalised data, for which the best fit lies along the 45° line.

---

[5] Irizarry R. A., Bolstad B. M., Collin F., Cope L. M., Hobbs B. and Speed T. P. 2003. 'Summaries of affymetric genechip probe level data.' *Nucleic Acids Research* 21: e15.

[6] Ritchie M. E., Silver J., Oshlack A., Holmes M., Diyagama D., Holloway A. and Smyth G. K. 2003. 'A comparison of background correction methods for two-colour microarrays.' *Bioinformatics* 23: 2700–2707.

[7] A log scale is used as it is a transformation that is symmetric around the line of no fold change. In the absence of this transformation, a two-fold and a two-hundred fold reduction in fluorescence signals both lie between 0 and 1, whereas a corresponding up-regulation will be more widely spaced. On a logarithmic (base = 2) scale, a two-fold up-regulation corresponds to a value of +1, with a corresponding down-regulation represented by −1; corresponding values for a 200-fold change are +7.6 and −7.6 respectively.

A second method, also uses linear regression, but for a different scatter plot, routinely referred to as the M–A plot, where $M = \log_2 R - \log_2 G$ and $A = (\log_2 R) + (\log_2 G)/2$ are plotted with M on the y-axis and A on the x-axis. Ideally, this plot should be represented by $M = 0$ as the best fit line. However, as with the $\log_2 G$ v. $\log_2 R$ scatter plot, this is generally not the case, and can be corrected using linear regression (Fig. 3.3 b).

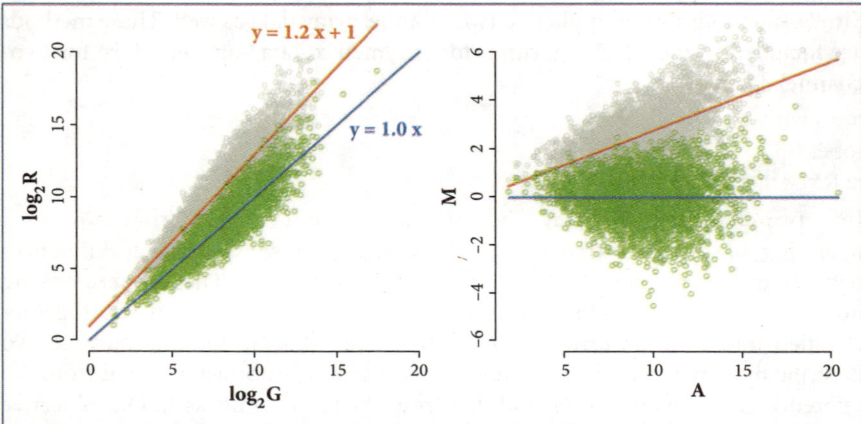

Figure 3.3 *Methods for within-array normalisation of two-colour microarray data.* (a) R-G plot: The grey dots show a scatter plot between the log-transformed signals from the two channels before normalisation, with the red line being its best-fit line. This clearly deviates from the 45° line, indicating systematic differences between the signals from the two channels. The green dots show the scatter plot after $\log_2 R$ has been adjusted, with the blue line being the 45° best-fit line. These are based on simulated data. (b) M-A plot, graphing the log-normalised fold change (M) between the two channels against the average log-normalised signal (A) of the two channels, for the same data shown in (a). Again, the grey dots and the red line show the data before normalisation, and the green dots and blue line, after normalisation.

The problem with using linear regression is that the relationships explored above are generally not linear. Therefore, a more inclusive method of adjusting deviations from the ideal expectation is to use local regression methods such as LOESS, which performs many local regressions in overlapping windows and then stitches them together into a smooth curve. On the M–A plot, the normalised log ratio M' is calculated as the difference between the raw M and the fitted value on the regression curve.

Finally, there is the possibility that there are systematic local artifacts on the microarray. For example, the array may not be perfectly horizontal and since different lasers have different depths of focus, the extent to which each laser is focused on the array might differ across spatially-separated regions (referred to as *grids*) on the microarray. This can be corrected for by applying the regression analysis separately over different grids on the microarray.

To summarise, the problem of within-array normalisation arises only for two-channel arrays, and is not an issue with one-channel systems such as Affymetrix arrays.

However, both one-and two-channel arrays face the problem that hybridisation conditions may differ slightly from array to array and so the overall brightness may be different across arrays. The overall distributions of signals across arrays should be brought to the same scale before meaningful comparisons can be made. One means by which this could be achieved is quantile normalisation, which results in arrays with the same distribution of signals, but with differences in the ranking of probes on the basis of their fluorescence signals (Fig. 3.4). This is implemented in the RMA algorithm, available in Bioconductor.

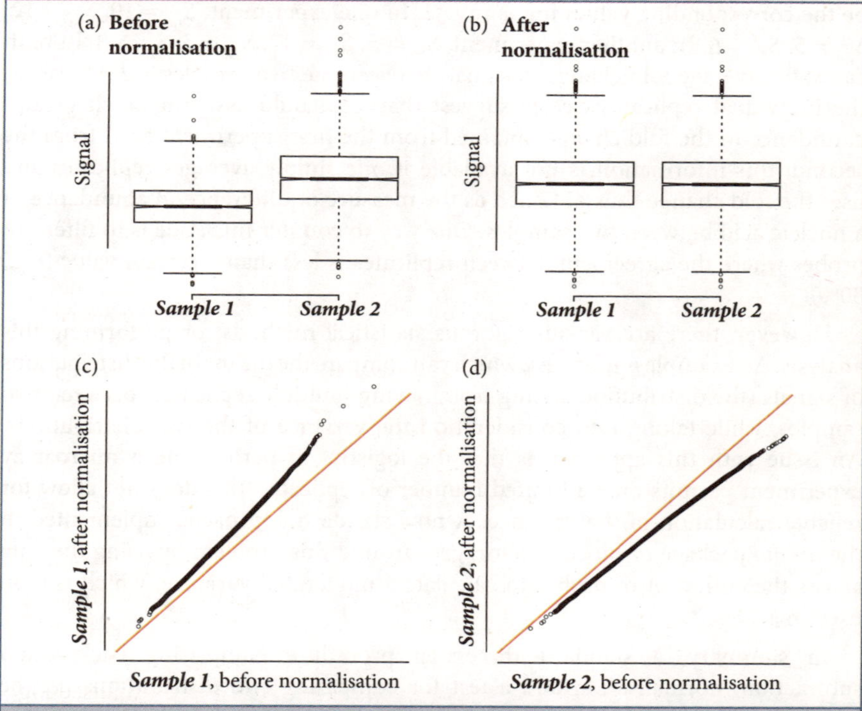

Figure 3.4 *Quantile normalisation.* (a) Distribution of signal for two samples – one brighter on average than the other – before normalisation; (b) distribution of signal strength for the two samples after quantile normalisation. (c and d) Scatter plots of the signal before (x-axis) and after (y-axis) normalisation, illustrating the corrections applied; the red diagonal is the 45° line, which represents no correction.

Finally, it is noted here that these methods operate under the assumption that for a majority of genes/probes, there is little difference between samples in their fluorescence signals. In cases where this is violated, a different experimental design is required, and the reader is directed to the primary literature for a reference.[8]

3.3.3 Differences in signal from the same probe between two samples

A simple means of testing whether a probe produces significantly different signals in two different samples is to calculate the average signal between multiple replicates for the same sample, and to test whether the difference in this average value between the two samples is greater than a pre-determined threshold (e.g., two-fold). However, this does not take into account variability among replicates from the same sample. For example, let $S_{1,a}$ and $S_{1,b}$ be the background-corrected and normalised signal for two replicates *a* and *b* for sample 1; and $S_{2,a}$ and $S_{2,b}$ be the corresponding values for sample 2. In one experiment, $S_{1,a} = 10$, $S_{1,b} = 12$; $S_{2,a} = 5$, $S_{2,b} = 6$. In another experiment, $S_{1,a} = 5$, $S_{1,b} = 17$, $S_{2,a} = 3$, $S_{2,b} = 8$. In both cases, the average fold change in signals between the two samples is 2. However, the individual replicates clearly suggest that one could assign a much greater confidence to the fold change obtained from the first experiment than from the second; this information is not available if one simply averages replicates and uses the fold change thus obtained as the measure of difference in abundance of a nucleic acid between two samples. One way to counter this issue is to filter out probes where the agreement between replicates is less than a certain value (e.g., 80%).

However, there are various rigorous statistical methods for performing this analysis. An example is the t-test, which can compare the means of the distributions of signals (the distribution arising from having multiple replicates) between two samples, while taking into consideration the variance of the two distributions. An issue with this approach is that the logistics of performing a microarray experiment permits only a limited number of replicates–this does not allow for reliable calculation of the variance. A now standard approach, implemented in the *limma* package of Bioconductor, gets around this problem by using the data across the entire set of probes to calculate a moderated variance, which is then used in a t-test.[9]

In summary, a standard three-step procedure comprising background subtraction, normalisation and a test for comparing two distributions can be

---

8 van de Peppel J., Kemmeren P., van Bakel H., Radonjic M., van Leenen D. and Holstege F. C. 2003. 'Monitoring global messenger RNA changes in externally controlled microarray experiments.' *EMBO Reports* 4: 387–93.

9 Smyth. 2004. 'Linear models and empirical Bayes methods for assessing differential expression in microarray experiments.' *Statistical Applications in Genetics and Molecular Biology* 3, 1: 3.

used to identify nucleic acid sequences whose abundances vary significantly between two samples. These approaches are applicable to both CGH–which will be discussed in this chapter–and measurement of gene expression changes, presented in Chapter 5. Specific analysis methods, such as *GACK*, have been developed for analysis of CGH microarray data–these use simple methods for identifying cutoffs within a distribution of fold changes (on a $\log_2$ scale), which help define absent and divergent genes.[10]

3.4 Comparative genome hybridisation and bacterial phylogenomics

Microarrays can be used to probe the gene content of bacterial genomes, as long as there is a suitable reference genome available, the sequence of which can be used to design probes for the microarray. Let us assume that there is an organism *A* whose genome sequence is available. Let us further assume that a virulent pathogen *B* emerges, which we know by various approaches to be a relative of *A*. We wish to catalogue, roughly, the variations between *A* and *B* on a genomic scale. We are unable to pursue this using the ideal approach of completely sequencing the genome of *B* and then performing a comparative genomic analysis to discover the variations, relative to *A*, that it harbours. We choose to go the microarray way (Fig. 3.5). Depending on how the probes have been designed, we could answer different types of questions. If short oligonucleotides were used as probes, for example in a genome tiling format, one could in principle detect small variations including SNPs that are present within the region probed by the oligonucleotide. We can refer back to the idea of a mismatch probe, used in Affymetrix arrays to facilitate background subtraction, which provides a useful conceptual frame work for understanding this: The fluorescence signal obtained from a probe will depend on the degree of variation between the probe and the hybridised sample, as well as the position of the variant base(s). However, the exact SNP or its exact position cannot be determined. Alternatively, one could use cDNA arrays in which entire ORF sequences are used as probes. Such a platform would help us decide whether an ORF is present in the test organism; this may not always be distinguishable from ORFs which are present, but too divergent in DNA sequence to hybridise well with the probe. Note however that, irrespective of the array design, a microarray experiment will not permit identification of sequences present only in *B*, but not in *A*. In other words, variant discovery using a microarray is restricted to sequences that are present on the array as probes. Therefore, the choice of array

10 Kim C. C., Joyce E. A., Chan K. and Falkow S. 2002. 'Improved analytical methods for microarray-based genome-composition analysis.' *Genome Biology* 3: RESEARCH0065.

design, as well as the set of probes used, will determine the nature of questions that can be asked of a CGH experiment.

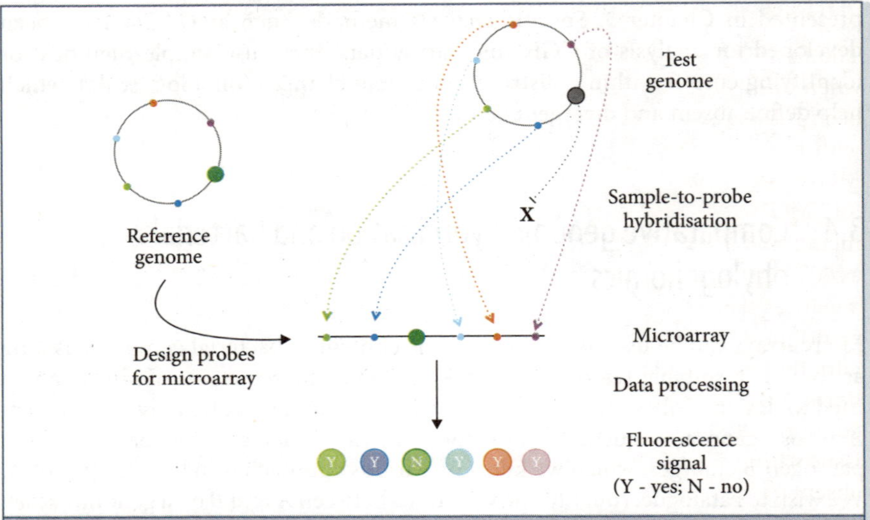

Figure 3.5 *A schematic representation of a CGH experiment.* In a CGH experiment, a microarray is designed based on the genome of an organism for which sequence data are available. Genomic DNA from a related organism is hybridized to this microarray to identify regions, which are absent or divergent in the sample relative to the sequence probed on the microarray. The filled circles on the circular chromosomes represent genes. The dark green gene (large filled circle on the reference genome), present in the reference genome but not in the test genome will be called as absent. The presence of the grey gene (large filled circle on the test genome), encoded by the test genome but not by the reference genome, cannot be discovered.

3.5 Case studies

In the rest of this chapter, we will discuss CGH studies in three different bacterial species: (a) *Escherichia coli*; (b) *Staphylococcus aureus*; and (c) *Helicobacter pylori*, to present aspects of experiment design as well as to illustrate concepts in bacterial gene content that have emerged. Note that high density tiling microarrays have been used to study recombination events in yeast at high resolution; these are not discussed here, but the reader is directed to the original publication.[11] Note also that this survey is not exhaustive, and only a few of the many relevant studies are

[11] Mancera E., Bourgon R., Brozzi A., Huber W. and Steinmetz L. M. 2008. 'High-resolution mapping of meiotic crossovers and noncrossovers in yeast.' *Nature* 454: 479–85.

covered.

3.5.1 Comparative genome hybridisation studies of *Escherichia coli*

The first CGH study of *E. coli* was published–by Ochman and Jones–at a time when only one *E. coli* genome–that of K12 MG1655 (see previous chapter)–had been fully sequenced.[12] In this study, the array was a nylon membrane spotted with the sequences of nearly 4,300 ORFs predicted in the fully sequenced K12 MG1655 genome. ^{32}P-labelled genomic DNA from five *E. coli* strains including K12 MG1655 itself and the very closely related K12 W3110 were hybridised to this membrane. The strains included in this study were reported to differ by as much as 800 kb in chromosome length. Using a method that was stringent in identifying a gene as absent in a particular strain of *E. coli* (no visible sign of hybridisation at all), the authors found that nearly 3,800 genes were common to all strains. As admitted by the authors, the number is likely to be an overestimate. This is in fact borne out by later comparative genomic surveys, which placed the number of genes common to five strains of *E. coli* at an average value of ~3,100–3,200. A second issue with this study was that the readout was binary–either a gene was present or absent. As a result, copy number variations, for example of transposable elements, could not be measured. Besides describing certain characteristics of genes with variable occurrence across the strains studied here (some gene functions are mentioned below, in the context of a more recent study), the authors of this work attempted to trace their evolutionary histories. This was possible because a reliable phylogenetic tree of these strains of *E. coli* was already available. A gene that is present in MG1655 could be absent in any of the other strains as a result of one of the following processes:

1. a gene might be present in the common ancestor of all these strains but subsequently lost in a non-MG1655 lineage (*deletion/gene loss* event);
2. a gene could be absent in the common ancestor, but gained subsequently somewhere along the branch leading to MG1655, but not in other lineages (*insertion/gene gain* event).

These events could be inferred on the basis of sequence alignments of MG1655 genes with those in *Salmonella* gene databases. Using this approach, the authors were able to explain the occurrence of all the genes included in their analysis by means of 30 deletion and 37 insertion events (Fig. 3.6 a), with loci classified under deletion events being shorter than those under insertion events. As pointed out in the previous chapter, horizontally-acquired genes generally have a base composition that is atypical of the rest of the genome. In this work, the authors

[12] Ochman and Jones. 2000. 'Evolutionary dynamics of full genome content in *Escherichia coli*.' *EMBO Journal* 19: 6637–6643.

observed that both deletion and insertion events covered genes with atypical base composition; this was more apparent for insertion events than deletions. The latter observation was justified by the argument that both recently acquired but subsequently useless, and ancestral genes could be deleted; thus, the sequence composition of deleted genes would be a hybrid of ancestral and recently-acquired genes.

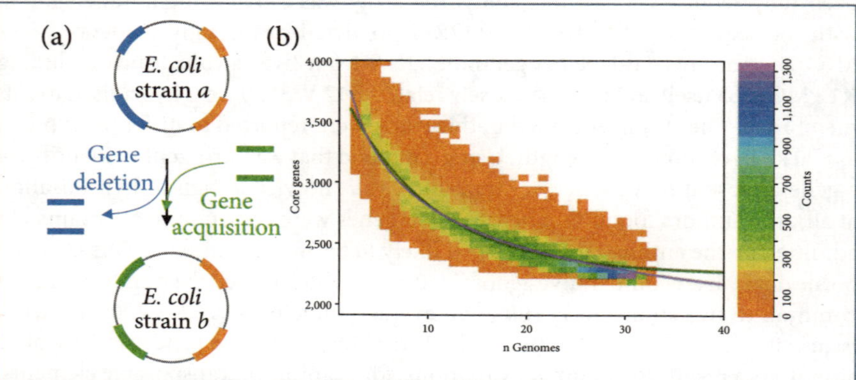

Figure 3.6 *Evolution of E. coli gene content.* (a) In this example, the hypothetical *E. coli* strain *a* is the immediate ancestor of *E. coli* strain *b*. The figure is similar to Fig. 2.7a in Chapter 2, but more realistic in showing the evolution of one *E. coli* strain from another by a combination of gene deletion and gene acquisition. Islands which are lost by strain *a* are marked in blue, and those gained by strain *b* are in green; red regions mark non-core elements which are common to both *a* and *b*, but not present in other *E. coli* strains. This is based on the work by Ochman and Jones showing that gene deletion and gene acquisition events are almost equally common in *E. coli* evolution. (b) This shows a two-dimensional density plot of the change in the size of the core genome of *E. coli*, as defined by the number of genes common to all strains under consideration, with the addition of every new *E. coli* genome. To quote from the original source (see below) of this figure, 'The plot illustrates the number of *E. coli* core genes for $n = 2, ..., 32$ genomes based on a maximum of 3,200 random combinations of genomes for each n. The density colors reflect the count of combinations giving rise to a certain number of core genes; that is, for $n = 3$, genome number 3 is compared to genomes 1 and 2, and the number of core genes is the number of genome 3 genes conserved in genomes 1 and 2. The green line is the fit to the exponential decay function …, and the red line is our proposed fit to a slightly modified decay function'. This figure panel is reproduced under the Creative Commons Attribution license from Willenbrock, Hallin, Wassenaar and Ussery. 2007. 'Characterization of probiotic *Escherichia coli* isolates with a novel pan-genome microarray.' *Genome Biology* 8: R267. © Willenbrock et al. 2007; licensee Biomed Central.

In a second major CGH study, a group from Japan probed the gene contents of 22 pathogenic strains of *E. coli* and *Shigella*,[13] using a microarray spotted with a little under 4,100 ORFs encoded in the genome of *E. coli* K12 W3110.[14] This microarray covered nearly 93% of the ORFs predicted in this reference genome. A small number of spots which showed low signal–compared to control spots derived from a human gene–when hybridised to genomic DNA from the reference organism itself (self-hybridisation) were excluded from further analysis. The availability of additional completely sequenced *E. coli* genomes allowed the authors to benchmark their microarray and identify a threshold for defining genes absent in the hybridised sample. This was done as follows. A comparative genomic analysis between *E. coli* K12 W3110 and *E. coli* O157: H7 SAKAI was performed using their complete genome sequences and annotations. Among the genes with valid probes on the microarray, 3,528 were considered as homologous in both genomes. From the microarray experiment, it was found that the presence of 3,525 of the above genes was correctly identified. This was based on $\log_2$ ratio (*test* – *reference* < = –1 for absent genes) on a two-colour array where the *reference* was W3110 DNA itself and the *test* sample was genomic DNA from O157: H7. Similarly, of the 345 genes that were defined as absent in O157: H7 but present in W3110 based on the comparative genomic sequence analysis, the absence of 344 genes was correctly determined by the microarray. This comparative sequence analysis-guided method of identifying cutoffs for defining present/absent genes in a CGH experiment was reported to be superior to *de-novo* definition of cutoffs–based entirely on the microarray data as exemplified by the GACK procedure–at least for this array design and experimental procedure. Of the <4,100 ORFs spotted on the microarray, >1,400 were considered as absent in at least one of the strains used in this study. Therefore, the conserved core was set to <2,600 genes (with an additional ~200 genes added, to account for those that were not probed by the microarray), which is in reasonable agreement with the number arrived at by a comparative genomic analysis of 17 *E. coli* isolates covering multiple pathovars (see previous chapter). Analysis of gene annotations showed that genes involved in fundamental cellular processes such as replication and translation were rarely absent in any strain. However, larger proportions of genes involved in carbohydrate transport and metabolism, cell surface structures such as the O-antigen and fimbriae, and regulatory proteins were divergent across strains. This reflects differences among strains of *E. coli* in their nutrient utilisation,

[13] Note that it is widely accepted that the separation of *E. coli* and *Shigella* into two distinct genera is phylogenetically invalid and only of historical relevance; therefore, many studies include *Shigella* species in their analysis of *E. coli* genomes.

[14] Fukiya S., Mizoguchi H., Tobe T. and Mori H. 2004. 'Extensive genomic diversity in pathogenic *Escherichia coli* and *Shigella* strains revealed by comparative genomic hybridisation microarray.' *Journal of Bacteriology* 186: 3911–3921.

antigenic potential and environment sensing abilities. Finally, the above work performed a simple clustering analysis of *E. coli* strains on the basis of the presence or absence of genes.[15] Briefly, a matrix in which each column represented a strain and each row a gene was constructed. Each entry in this matrix was either a '1' or a '0', indicating the presence or absence of a gene in a genome. Clustering of this matrix showed that this simple approach to phylogenomic analysis could give a reasonable picture of lineage diversification of *E. coli*.

The two studies discussed above accessed genes that were present in a single *E. coli* genome, which was probed by the microarray–a major handicap in studying such a genetically diverse species complex. This issue has been addressed using pan-genome arrays covering genes across several strains of *E. coli*. We will discuss two such studies here, the first based on a commercially available microarray covering genes from multiple *E. coli* strains and the second discussing aspects of array design from first principles. In the first study, Wick and co-workers used CGH to study gene content variation within the O157: H7 lineage of *E. coli*.[16] They used a commercial microarray, which contained <6,200 50-mer probes covering three *E. coli* genomes: Those of K12 MG1655 and two O157: H7 strains (EDL933 and SAKAI). The array also had <70 control probes interrogating the genome of the plant *Arabidopsis thaliana*. The authors benchmarked the specificity of their array against the fully-sequenced genomes of K12 MG1655 and O157: H7 SAKAI and found that only <5,700 probes had unique targets (non-repetitive regions). They also detected some inconsistencies in the probe annotations provided by the manufacturer and corrected these. This done, they used a two-colour hybridisation strategy with genomic DNA from O157: H7 SAKAI as the reference. To make presence/absence calls, they used the GACK algorithm and found that even at a sequence similarity of 94%, there was a 50% chance that a sequence would be called absent. This number, obtained using oligonucleotide probes, was not much worse than those obtained using cDNA probes in other organisms. Thus, an important outcome of this analysis is that CGH studies could at times be over-sensitive in calling absent/divergent genes. Though this may be different across different experimental systems and strategies, benchmarking against a few reference genomes is important, as this provides estimates of error rates for that study. The above study showed that even within the same lineage of *E. coli* (that leading to O157: H7), over 11% of genes were absent in at least one strain. These

---

15 This is similar to the example shown for comparative genome sequence analysis in Figure 2.14, and later in this chapter for *Helicobacter pylori* in Figure 3.9. This is also analogous to the phylogenetic profile used in non-homology-based methods of gene annotation, discussed in the previous chapter.

16 Wick L. M., Qi W., Lacher D. W. and Whittam T. S. 2005. 'Evolution of genomic content in the stepwise emergence of *Escherichia coli* O157: H7.' *Journal of Bacteriology* 187: 1783–1791.

variable elements included several prophage genes, and those encoding virulence factors and O-antigen biosynthetic enzymes.

The above study was again limited in that probes were designed against only two *E. coli* pathovars. However, with more genome sequences becoming available, it became possible to design probes covering a large number of related genomes. David Ussery's group in Denmark has been pioneering such efforts for *E. coli*. In one such study, Willenbrock and colleagues performed a full comparative genomic analysis of 32 strains of *E. coli* and *Shigella*.[17] The core genome size was estimated to cover between 2,050 and 2,250 genes depending on sequence comparison parameters (Fig. 3.6 b), and the pan-genome to cover between 9,500 and 12,000 genes with every new *E. coli* genome expected to uncover ~80 new genes in *E. coli*'s genetic arsenal. This comparative analysis of genome sequences provided a framework for designing a comprehensive microarray for *E. coli* CGH experiments. The probe design strategy was the following (Fig. 3.7). The same gene prediction algorithm (EasyGene)[18] was applied to all the 32 *E. coli* genomes in order to standardise gene annotations. Genes from across all these strains were then grouped together based on certain sequence similarity thresholds. Those falling within each group were aligned using the multiple sequence alignment program ClustalW. A consensus sequence was derived for each sequence group from these alignments. Probes were designed against this consensus using the program OligoWiz,[19] which ensures that (a) probe cross-hybridisation is minimised; (b) probe melting temperature distribution lies within a narrow range; (c) probes fold minimally to maximise sensitivity; (d) probes avoid low complexity sequences. In addition, it was ensured that any mismatch between a probe and its hybridising sequence from the genome was more likely to be closer to the edges of the probe, so that it affects hybridisation minimally. This procedure resulted in 305,000 55–60-mer probes covering nearly 12,000 gene groups. Next, these probes were aligned to all the gene sequences from the 32 genomes, and those that matched against multiple sequence groups, and those that did not match against every member of a single gene group were filtered out. This additional filtering step, which resulted in a final set of <225,000 probes covering ~9,250 gene groups, was demonstrated to be critical in minimising non-specific hybridisation signals. The pan-genome microarray was then used to interrogate the gene content of several non-pathogenic probiotic *E. coli* strains; the readers are referred to the original publication for further details.

---

[17] Willenbrock H., Hallin P. F., Wassenaar T. M. and Ussery D. W. 2007. 'Characterization of probiotic *Escherichia coli* isolates with a novel pan-genome microarray.' *Genome Biology* 8: R267.

[18] Larsen and Krogh. 2003. 'EasyGene-a prokaryotic gene finder that ranks ORFs by statistical significance.' *BMC Bioinformatics* 4: 21.

[19] Wernersson and Nielson. 2005. 'OligoWiz 2.0-integrating sequence feature annotation into the design of microarray probes.' *Nucleic Acids Research* 33: W611–15.

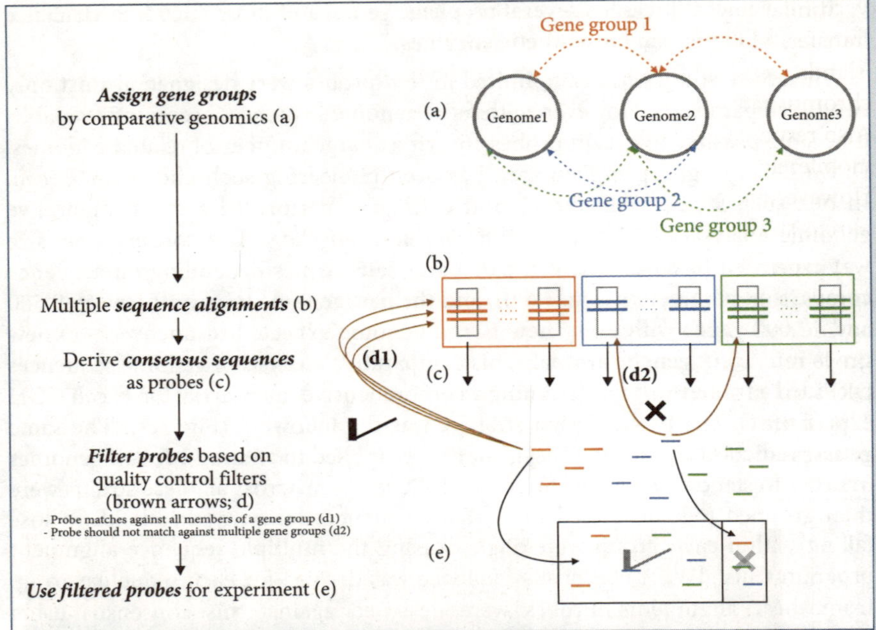

Figure 3.7 *Designing a multi-genome oligonucleotide probe for CGH experiments.* A schematic representation of a procedure for designing a multi-genome oligonucleotide microarray for CGH experiments. The protocol shown is a simplified version of that used by Willenbrock and colleagues while designing a microarray for *E. coli* CGH experiments. The method is described in the main text. The procedure for deciding on sequences for PCR product microarrays is conceptually similar in that this should also result in probes that align to all orthologs of a gene across the reference strains, and should not cross-hybridise with other sequences (even from the same sequence family); approaches taken in this direction are discussed in the context of studies of *Staphylococcus aureus* and *Helicobacter pylori.*

3.5.2 Comparative genome hybridisation studies of *Staphylococcus aureus*

Staphylococcus aureus is a major human pathogen, responsible for various mild skin infections as well as life-threatening conditions causing septicemia, endocarditis, necrotising pneumonia and toxic shock syndrome. It is a common source of hospital-acquired infections, with community-associated versions being increasingly reported. A major cause of concern with *S. aureus* infections is the ability of the pathogen to resist antibiotics, most famously methicillin (MRSA, for methicillin resistant *Staphylococcus aureus*).

Similar to *E. coli*, genome evolution in *S. aureus* is driven by horizontal gene transfer. Many mobile genetic elements–including bacteriophages, pathogenicity islands and antibiotic resistance-conferring SCC (Staphylococcus cassette chromosome) are responsible for disease and the organism's success as a pathogen. Also responsible are various 'core variable' (CV) genes, which are not carried on mobile elements, but appear to be variably gained or lost in different lineages. However, unlike in *E. coli*, there are no large variations in genome size among *S. aureus* strains, with most genomes being around 2.8 Mb.

Fitzgerald and co-workers reported–in 2001–an early CGH study of 36 *S. aureus* isolates recovered from different human diseases and host types (human, bovine and ovine).[20] The microarray was designed to cover >90% of all ORFs of a single reference strain (named COL). The study found that ~2,200 of the ~2,800 ORFs on the micorarray were common to all the 36 isolates used. The variable 22% of the ORFs were located in contiguous regions ranging in size from 3–50kb, as assessed from the genome of the reference strain COL (Fig. 3.8). Of the 18

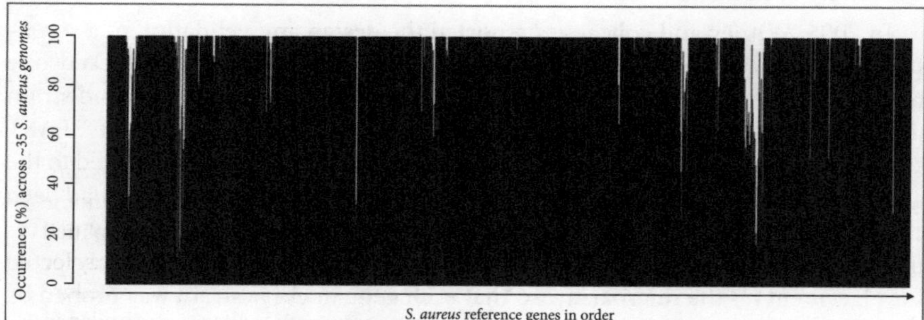

Figure 3.8 *Gene conservation in Staphylococcus aureus*. The figure shows data from a CGH experiment, by Fitzgerald and co-workers (Fitzgerald et al. 2001), covering >35 *S. aureus* strains. All genes from the reference genome are lined up along the *x*-axis and for each gene, the percentage of *S. aureus* strains in which it was deemed to be present plotted as bars on the *y*-axis. White patches show contiguous stretches of genes which are absent in some of the strains.

contiguous regions of divergence described in this study, as many as 10 encoded annotated virulence factors and mediators of antibiotic resistance. Similar to analysis performed in some *E. coli* CGH studies, clustering of a presence/absence matrix of genes across the 36 isolates was performed. This allowed the authors to

[20] Fitzgerald J. R., Sturdevant D. E., Mackie S. M., Gill S. R. and Musser J. M. 2001. 'Evolutionary genomics of Staphylococcus aureus: Insights into the origin of methicillin-resistant strains and the toxic shock syndrome epidemic.' *Proceedings of the National Academy of Sciences USA* 98: 8821–8826.

observe that the gene for methicillin resistance (*mec*, responsible for MRSA) was sporadically present, and in very different clusters/lineages. They then stated that MRSA strains have evolved multiple times independently, as opposed to being derived from a single common ancestor.

As with several *E. coli* CGH studies, the above study had the disadvantage of being based on a microarray derived from a single reference genome, which is sub-optimal especially for organisms with a great diversity of mobile genetic elements and therefore, gene content. Several studies have made use of multi-strain microarrays, covering ORFs from several completely sequenced *S. aureus* genomes. Similar to the Willenbrock study for *E. coli*, oligonucleotide probes covering six *S. aureus* genomes have been designed and used for CGH studies.[21] In addition, cDNA arrays, using PCR amplicons of ORFs covering seven fully-sequenced *S. aureus* genomes, have been developed and used extensively. In the rest of this section, we will discuss the design of this cDNA array and present some findings obtained from its use. This array was developed by Jodi Lindsay from the University of London.

In 2005, Witney and colleagues reported the design and validation of a seven-strain microarray for *S. aureus.*[22] To design the array, the authors picked one reference strain and added all its ORF sequences to a 'gene pool'. A second strain was picked and all genes unique to this strain were selected based on BLAST searches against the current gene pool. These unique ORFs were then added to the gene pool. The process was iterated until ORFs from all the seven strains had been considered and relevant subsets added to the gene pool. Regions of the sequences in the gene pool were amplified (~600 bp as optimal length) and amplicons selected for placement on the microarray, so that each gene in every strain was probed by the array and each selected PCR product had minimal similarity to paralogous genes. Finally, ~3,600 genes were covered by the array. For the hybridisation experiments, genomic DNA from one strain–the strain used to first populate the gene pool–was used as reference. Thus, log ratios in the signal between the test strain and the reference would identify genes present in the former, but not in the reference, as long as these were present in one of the seven fully-sequenced *S. aureus* genomes and therefore, probed by the microarray. This is in addition to the standard application of identifying genes present in the reference genome but not in the test DNA. Using a particular method of defining a cutoff for identifying

---

[21] Dunman P. M., Mounts W., McAleese F., Immermann F., Macapagal D., Marsilio E., McDougal L., Tenover F. C., Bradford P. A., Petersen P. J., Projan S. J. and Murphy E. 2004. 'Uses of Staphylococcus aureus genechips in genotyping and genetic composition analysis.' *Journal of Clinical Microbiology* 42: 4275–4283.

[22] Witney A. A., Marsden G. L., Holden M. T., Stabler R. A., Husain S. E., Vass J. K., Butcher P. D., Hinds J. and Lindsay J. A. 2005. 'Design, validation and application of a seven-strain Staphylococcus aureus PCR product microarray for comparative genomics.' *Applied and Environmental Microbiology* 71: 7504–7514.

present and absent genes, it was found that the array correctly identified ~97% of genes as present/absent. A significant proportion of the wrong identifications could be explained by the presence of the relevant gene on a plasmid or a transposon (copy number effects), or gene sequence redundancy and very small genes producing weak signals.

Lindsay et al. used the above-described seven-strain microarray to probe the gene content of 161 *S. aureus* isolates covering both disease-associated and asymptomatic carrier varieties.[23] A gene occurrence (presence/absence) matrix was constructed based on >700 core-variable genes (see above), which included many surface associated genes such as those involved in capsular biosynthesis and proteins that bind to host tissues. Using this matrix, the authors compared lineages/clusters obtained from this genome-scale analysis with those obtained from the popular multi-locus sequence typing (MLST) method, where lineages are obtained on the basis of polymorphisms in seven housekeeping genes. The analysis showed that the core-variable gene content reflected the MLST tree at least at the level of clonal complexes,[24] though not necessarily at the level of the sequence type.[25] Mobile genetic elements, not included in the above-described clustering analysis, were found to be variable, sometimes even within lineages. Such variations included conservation of short fragments of a mobile genetic element, but not the rest of the genes within the element. For example, genes for a virulence factor called PVL (Panton–Valentine leukocidin), a toxin, were carried on a certain phage called Φ2 in one strain; however, none of the other strains which carried the genes for the PVL toxin were positive for the rest of the Φ2 phage suggesting that this virulence factor was encoded in unrelated phages in different strains. Finally, the study was unable to find statistically significant differences between invasive and carrier strains in their gene content.

In the final study that we will discuss in this section, Sung and colleagues used the seven-strain *S. aureus* microarray to probe the gene content variation between human- and animal-associated *S. aureus*.[26] For this, they used the 161 *S. aureus* human isolates used in the above study in addition to 56 isolates which had infected cows, horses, goats, sheep and camel. Clustering of the presence/absence matrix for the core-variable genes (see above) showed that animal and

---

[23] Lindsay J. A., Moore C. E., Day N. P., Peacock S. J., Witney A. A., Stabler R. A., Husain S. E, Butcher P. D. and Hinds J. 2006. 'Microarrays reveal that each of the ten dominant lineages of Staphylococcus aureus has a unique combination of surface-associated and regulatory genes.' *Journal of Bacteriology* 188: 669–75.

[24] Isolates with the same polymorphism in at least fie of the seven housekeeping genes are grouped within the same clonal complex.

[25] Isolates with identical polymorphisms across all the seven genes are grouped within the same sequence type.

[26] Sung J. M., Lloyd D. H. and Lindsay J. A. 2008. '*Staphylococcus aureus* host specificity: Comparative genomics of human versus animal isolates by multi-strain microarray.' *Microbiology* 154: 1949–1959.

human isolates did not separate into clearly distinct clades. The ten lineages in which the animal isolates were found clustered intermittently among the human-associated lineages. Further, only four of these animal-associated lineages exlusively contained animal-derived isolates. Evidence was found that some of the horse-associated lineages could have human origins. However, there were several key differences between human-and animal-associated isolates. Antibiotic resistance markers were rarely found in animal isolates, reflecting a general lack of SCC, plasmids and transposons in these organisms. Though bacteriophage and pathogenicity islands were found in many isolates, certain virulence factors were common in human-associated isolates but rare in those that had infected animals. However, as noted by the authors, the seven-strain microarray had been designed on the basis of human-associated strains and therefore, genes unique to animal-associated lineages could not be detected. This once again underlines the influence of microarray design and probe content on how the data could be interpreted.

3.5.3 Comparative genome hybridisation studies of *Helicobacter pylori*

Helicobacter pylori is a human pathogen which colonises the stomach and causes diseases ranging from gastritis to gastric cancer. Its genome, selectively described in the previous chapter, was sequenced as part of one of the earlier bacterial genome sequencing efforts. *H. pylori* has no known environmental reservoir and is vertically transmitted in families. It is typically acquired during childhood and carried by the individual through his/her lifetime. In fact, *H. pylori* genetic diversity, to some extent, is a proxy for human genetic variation. In developing countries, up to 80% of the population might be carriers of *H. pylori*, most asymptomatically. Mixed colonisation of the same individual by genetically distinct populations of this bacterium is also possible. This could promote recombination as a consequence of the organism's natural competence and high recombination rates. This, combined with the potential of this organism to generate antigenic variation at contingency loci by slipped-strand mispairing (see previous chapter) leads to considerable genetic variation. However, for a long time, only two fully sequenced *H. pylori* genomes were available (strains 26695 and J99). Inspite of this, its genetic diversity meant that it became a popular system for a variety of CGH studies interrogating the gene content of several clinical isolates.

An early study by Salama and colleagues, from a group led by Stanley Falkow, established a PCR product microarray for *H. pylori*, which has been used subsequently in several CGH studies.[27] This array was based on ORF sequences from both the reference genomes of *H. pylori* available at that time. The strain 26695 was arbitrarily chosen as the primary reference and PCR amplicons from

[27] Salama N., Guillemin K., McDaniel T. K., Sherlock G., Tompkins L. and Falkow S. 2000. 'A whole-genome microarray reveals genetic diversity among Helicobacter pylori strains.' *Proceedings of the National Academy of Sciences USA* 97: 14668–14673.

all its ORFs spotted on the array. In addition, ORFs unique to J99, as identified by sequence analysis, were added to the array. The authors ensured that the PCR amplicons from a gene had no cross-homology with other genes, including members of the same sequence family, in the same genome. The final array design contained probes for 1,660 ORFs (~99% of all ORFs considered for amplification). By hybridising genomic DNA from the two reference strains to the array, the authors found that strain-specific genes could be detected with a sensitivity of 96% and a specificity of 98%, with a two-fold difference in signal acting as a threshold for identifying absent genes. The numbers were better for strain 26695 than for J99; this is not surprising given that the genomic DNA of 26695 served as the primary reference for the array design.

Using the above array, the authors investigated the gene content of 15 *H. pylori* clinical isolates, using a 1:1 mixture of 26695 and J99 genomic DNA as the reference in a two-colour hybridisation format. Of the >1,640 genes that could be reliably interrogated, ~1,280 were found to be common to all the 15 isolates (Fig. 3.9).

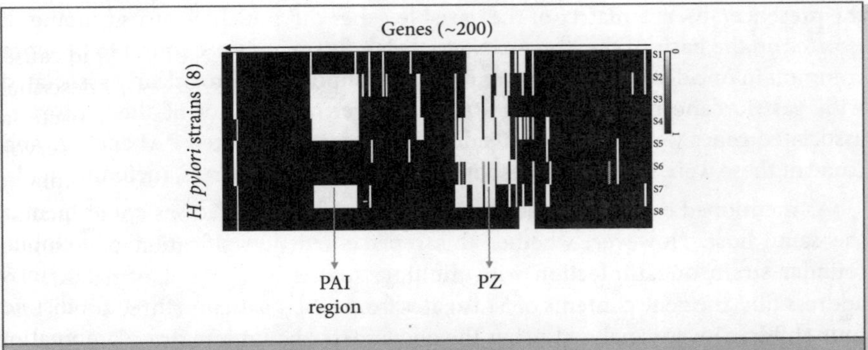

Figure 3.9 *Gene content of Helicobacter pylori genomes.* This is a heat map showing the occurrence of a set of ~200 genes (lined up in order along the horizontal) across 8 *H. pylori* genomes (vertical axis). Black indicates the presence of a gene in a genome, and white its absence. The CAG pathogenicity island (PAI) and a so-called plasticity zone (PZ) are marked. This is based on CGH data published by Salama and co-workers (Salama et al. 2000).

It was estimated that 12–18% of each genome was composed of strain-specific genes. As expected, the conserved genome encoded many essential metabolic and cellular functions, whereas a large number of the variable genes had no known functions. Among the variable genes were several outer membrane proteins and lipopolysaccharide biosynthetic genes. Interestingly, several members of restriction–modification systems were found to be variable, in agreement with more recent studies showing considerable variation between *H. pylori* strains in the architecture of their R–M systems.[28] Many of the variable genes were found in two large clusters: The first was a pathogenicity island (PAI) and the second was a plasticity zone (PZ). While the PAI was either present or absent as a cluster (with only a few exceptions), the PZ–which comprised many transposable elements–showed mosaicism with certain genes being present and others being absent (Fig. 3.9). A clustering analysis of the presence/absence matrix also identified certain chromosomally distal genes as being present/absent together with the PAI; these are candidates for novel virulence-associated factors. Traditionally, *H. pylori* strains had been typed based on the occurrence of the gene for a virulence factor CagA, encoded in the PAI. However, the clustering analysis showed, that PAI-lacking strains did not cluster together and away from those encoding the PAI.

In a second CGH study, the above-described array design was used to investigate gene content differences among isolates associated with three different disease outcomes–nonatrophic gastritis, duodenal ulcer and gastric cancer.[29] Among the 42 isolates studied in this work, >1,300 genes were found to be conserved with <350 genes being variable. Hierarchical clustering of these isolates based on the presence/absence matrix of the variable genes did not show any grouping of isolates on the basis of the disease. Nevertheless, 30 genes were found to be more common in one disease type than another. An important factor clearly associated with gastric cancer was the Cag PAI. However, a majority of these disease-associated genes were encoded outside the virulence-associated PAI and PZ, and some of these were involved in response to environmental stress, including acid.

As mentioned earlier, genetically distinct *H. pylori* populations could inhabit the same host. However, whether this represented diversification of a single founder strain or co-infection with multiple strains is a matter of debate. To address this, the gene contents of 51 isolates from seven patients (three adults and four children) were analysed using the above-described microarray.[30] A total of <1,300 genes were found to be conserved across these strains. Typing procedures had previously shown that clones from all but one patient were similar–the one exception had two distinct populations. This difference could be reproduced by

[28] (a) Lin L. F., Posfai J., Roberts R. J. and Kong H. 2001. 'Comparative genomics of the restriction-modification systems in Helicobacter pylori.' *Proceedings of the National Academy of Sciences USA* 98: 2740–2745; (b) Seshasayee A. S1., Singh P. and Krishna S. 2012. 'Context-dependent conservation of DNA methyltransferases in bacteria.' *Nucleic Acids Research* 40: 7066–7073.

[29] Romo-González C., Salama N. R., Burgeño-Ferreira J., Ponce-Castañeda V., Lazcano-Ponce E., Camorlinga-Ponce M. and Torres J. 2009. 'Differences in genome content among Helicobacter pylori isolates with gastritis, duodenal ulcer, or gastric cancer reveal novel disease-associated genes.' *Infection and Immunity* 77: 2201–2211.

[30] Salama N. R., Gonzalez-Valencia G., Deatherage B., Aviles-Jimenez F., Atherton J. C., Graham D. Y. and Torres J. 2007. 'Genetic analysis of Helicobacter pylori strain populations colonizing the stomach at different times postinfection.' *Journal of Bacteriology* 189: 3834–3845.

clustering the presence/absence matrix for the variable genes. *H. pylori* strains with similar typing footprints had ~20–70 variable genes among them. However, the two distinct clades from the one patient mentioned above had >110 variable genes, which included the PAI. Consistent with the fact that the two populations were isolated from spatially distant sites, sequence analysis of selected loci showed limited genetic exchange between them. This could be interpreted on the basis of the CGH data, which showed that the two populations differed from each other in the content of as many as nine restriction–modification systems. It was also reported that the children and the adults did not differ from each other in the divergence among the isolates they carried. Further, a volunteer who was experimentally infected with a *H. pylori* variant for a vacciniation trial, did not show any genetic diversification in his *H. pylori* population 15 and 90 days post-inoculation. These results do not support genetic diversification of a single founding strain within the host. Instead, evidence for co-infection was found with no suggestion of genetic exchange between the two infecting populations.

It must be noted here that though these experiments revealed patterns of gene loss/gain among *H. pylori* strains, they may be silent about point mutations and tandem repeat length variation, which are also common in *H. pylori*, with the latter elegantly explored by the genome sequencing projects themselves (see previous chapter).

Summary

- ✓ Microarrays are a major tool for studying gene content variation among closely related bacteria.
- ✓ Data produced by microarrays need to be processed computationally before meaningful conclusions can be drawn from them. These steps include background subtraction, normalisation and identifying difference in signal from a probe between two samples.
- ✓ The design of a microarray experiment (oligonucleotide tiling/PCR product; one-colour/two-colour) decides the kind of analysis pipelines one should adopt and the interpretations one can draw from an experiment.
- ✓ Multi-strain microarrays that include probes from different related genomes are a powerful tool for comparative genome hybridisation experiments. Their design requires careful thought.
- ✓ It is important to benchmark microarrays and analysis pipelines using genomic sequence comparison methods before deciding on their use.

Studying Bacterial Genomes using Next-Generation Sequencing

4

4.1 Introduction

In the previous chapter, we had emphasised the enormous interest in studying fine-scale genetic variation even among closely related bacteria. Given the cost and time involved in pursuing Sanger sequencing for a large number of genomes, microarrays came into vogue for determining variation in gene content among related organisms. However, microarrays suffered from the following limitations. All said and done, hybridisation signals from microarrays remain a poor proxy for direct sequence determination.[1] The exact nature and the extent of sequence divergence can not be determined. Second, the information that could be obtained depended on whether there were reference sequences for designing probes. Sequences that were unique to the genome sequence under investigation could not be determined. Therefore, there was a definite need for nucleic acid sequencing methods that could bypass the time and cost constraints of automated Sanger sequencing. In response to this need, there has been an explosion of *next-generation* or *deep* sequencing technologies in the last eight years or so. These technologies produce large volumes of sequence data in a short period of time and at a small fraction of the cost of a 'first-generation' automated Sanger sequencing experiment. They enable not just a reading of a DNA sequence (discussed in this chapter), but also permit a range of semi-quantitative applications typically associated with DNA microarrays. Popular next-generation sequencing systems are Illumina (also called Solexa), SOLiD (Applied Biosystems), 454 (Roche),

[1] Note however that in the 1980s, hybridisation of DNA to short oligonucleotide probes was proposed as a method for sequencing. This approach was a failure. However, the use of de Bruijn graphs for genome assembly, discussed briefly in this chapter, was first proposed as a method for assembling genomes from data generated by this sequencing-by-hybridisation technique (Pevzner. 1989. '1-Tuple DNA sequencing: Computer analysis.' *Journal of Biomolecular Structure and Dynamics* 7: 63–73).

Heliscope (Helicos Biosciences) and SMRT (Pacific Biosciences). Several other technologies including nanopore sequencing are in various stages of development and commercialisation. Benchtop sequencers, which are becoming increasingly available, make these technologies affordable to individual laboratories as opposed to core facilities.

In this chapter, we will provide a very brief overview of some general concepts underlying various next-generation sequencing technologies. This will be followed by an overview of the concepts underlying various computational procedures applicable to the use of these technologies for genome sequencing and re-sequencing. Following this, in line with the philosophy of this book, we will discuss specific investigations of bacterial genomes and learn some technical and biological lessons borne out of these studies. There may be a bias towards methods and studies based on the sequencing technology sold by the company Illumina;[2] this merely follows from the familiarity of this author with data produced by this technology, as well as the fact that Illumina has been the market leader in this domain, at least in terms of the number of peer-reviewed publications.

4.2 Next-generation sequencing technologies

The range of next-generation sequencing technologies can be classified based on at least two very broad parameters:

1 *Read length and sequencing depth*: The technologies of Illumina, SOLiD and Heliscope produce many millions of short (~100-mer) sequence reads. For example, a single run of Illumina sequencing[3] can produce up to 3 billion 100-mer reads, which translates to a 100-fold coverage of the human genome and 100,000-fold coverage of an average bacterial genome. On the other hand, the 454 pyrosequencing technology and the single-molecule real-time (SMRT) sequencing from Pacific Biosciences produce smaller numbers of considerably longer reads (>400-mer for 454 and 1,000 –3,000-mer for SMRT). Naively speaking, one could immediately state that the longer reads offered by the latter two technologies make them better-suited to *de-novo* genome sequencing based on the relative ease of sequence assembly; whereas, the great depth of an Illumina or a SOLiD sequencing run is ideal for semi-quantitative experiments such as gene expression

---

[2] This is not to belittle other technologies, which do offer their own advantages and disadvantages when compared to Illumina. For example, 454 pyrosequencing (Roche) produces much longer reads than Illumina, whereas SOLiD (Applied Biosystems) offers greater depth of sequencing. Different technologies have different sequencing error characteristics. Therefore, different applications may benefit more from the use of one technology than the other.

[3] On the HiSeq-1000 sequencer model.

measurements. However, the cost effectiveness of the shorter read technologies (at least in comparison to 454), plus the development of sophisticated algorithms for genome assembly from short reads, have made them immensely popular for both *de-novo* and reference-assisted genome characterisation, at least for the smaller genomes of bacteria.

2 *Sequencing approach*: Illumina, SOLiD and 454 sequencers use clonally amplified templates for sequencing. On the other hand, Heliscope and SMRT directly sequence single molecules of DNA (or RNA). The latter methods of sequencing, though not very common today, are likely to become more popular in the future, as these bypass the need for a PCR step which might introduce sequence biases, and can also do with relatively smaller quantities of the template nucleic acid. Some principles of template preparation and sequencing are described below. Detailed descriptions of these technologies are available in a review by Metzker.[4] The websites of companies offering these technologies also provide accessible graphical summaries of their approaches.

4.2.1 Template preparation strategies

The success of a sequencing technology depends on it being able to detect a signal, typically fluorescence, arising from a chemical reaction, namely the addition of a nucleotide to an extending DNA chain. Most detectors are unable to record signals arising from the addition of a nucleotide along a single template DNA molecule. Therefore, these technologies require clonal amplification of single template molecules by PCR before sequencing can be performed, so that each sequencing read represents a consensus derived from reactions occuring on the many clonal copies of the same template molecule (Fig. 4.1).

1 *Solid phase amplification*: The Illumina sequencing technology uses solid phase PCR amplification. Here, DNA fragments are ligated to adapters containing a universal priming site. These templates are then attached to a glass slide containing a high density of primers. The PCR produces 'clusters' of spatially adjacent amplicons, all being strand-specific copies of the original template. Each cluster produces single consensus sequence reads representing the template molecule. Each slide could contain over a billion clusters, thus producing high sequencing depth.

2 *Emulsion PCR*: Technologies such as 454 and SOLiD use what is called an emulsion PCR for the amplification step. Here, similar to the solid phase amplification strategy, the DNA to be sequenced is fragmented and then ligated to adapter sequences containing a universal priming site. These template fragments are then captured on primer-coated beads such that each bead

[4] Metzker. 2010. 'Sequencing technologies: The next generation.' *Nature Reviews Genetics* 11: 31–46.

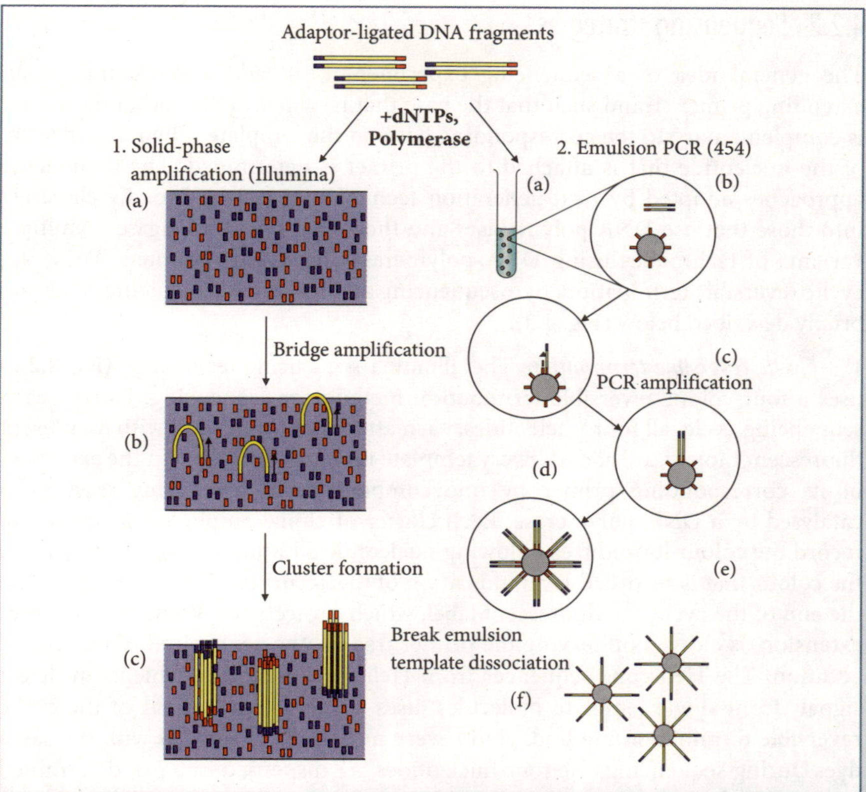

Figure 4.1 *Sample preparation strategies for next-generation sequencers.* This figure shows schematic representations of the solid-state amplification and emulsion PCR strategies used by next-generation sequencing technoloogies as part of their sample preparation pipelines. Illumina sequencers use solid-state amplification using primers embedded on a slide, which ultimately results in clonal sequence clusters; each cluster produces one consensus read sequence and thus the number of clusters that can be generated on the slide determines the depth of coverage. In contrast, 454 sequencers use emulsion PCR in which primers are coated on beads, which act as the substratum for the PCR reaction. The procedure is designed such that each bead will amplify a single sequence, and thus be equivalent to a single sequence read. Figure credit: Avantika Lal, National Centre for Biological Sciences, India.

contains one single-stranded template molecule. Following amplification of the template, the beads are chemically attached to glass slides or picotitre wells for the sequencing reaction to be carried out. Each bead produces one sequence read corresponding to the sequence of that single template molecule initially attached to the bead.

4.2.2 Sequencing strategies

The general idea of a sequencing experiment is to add a nucleotide to an extending primer strand such that the base that is added to the end of the strand is complementary to the corresponding base on the template. Then, the identity of the nucleotide that is attached to the primer is determined. The sequencing approaches adopted by next-generation technologies can be broadly classified into those that use DNA polymerases and those that use DNA ligases. Multiple variants of techniques using DNA polymerases have been adopted: These are cyclic reversible termination, pyrosequencing and real-time sequencing. They are briefly described below (Fig. 4.2).

1. *Cyclic reversible termination*: The Illumina sequencing technology (Fig. 4.2 a) uses a four-colour reversible termination method for sequencing. During each sequencing cycle, all four nucleotides, each attached at its 3′-end with a different fluorescent moiety, are added. Every template molecule will support the extension of its corresponding primer by one complementary base. This reaction is catalysed by a DNA polymerase. Each cluster of clonal amplicons is imaged to record the colour it produces following nucleotide addition. The homogeneity of the colour that is recorded is an indication of the accuracy of the base calling. At the end of the cycle, the fluorescent label, which protects the 3′-end from further extension, is cleaved off leaving the primer free for the next round of nucleotide addition. The Heliscope sequencer from Helicos Biosciences, which can detect signals from single template molecules, uses a one-colour variant of the cyclic reversible termination method. Here, every nucleotide is labelled with the same dye. During sequencing, the four nucleotides are dispensed in a pre-determined order, one after the other. After the addition of each nucleotide, templates emitting fluorescence are identified and the added nucleotide read as part of their sequence. Again, cleavage of the 3′-linked fluorescent label regenerates the end for further extension.

2. *Pyrosequencing*: Unlike other sequencing technologies, which make use of modified bases, pyrosequencing by 454 (Fig. 4.2 c) adds a single pre-determined unmodified nucleotide at every cycle. Template-containing beads to which the newly added nucleotide is attached, are identified by converting the local release of pyrophosphate into a light signal using a series of enzymatic steps.[5]

3. *Real-time sequencing*: The SMRT sequencing system by Pacific Biosciences uses a four-colour scheme for representing nucleotides. However, instead of pausing DNA synthesis to perform detection and/or cleavage reactions, the

[5] Note that the 454 sequencing technology is on the way out today. This is being replaced by the next generation of semiconductor sequencing technologies, under the name "Ion". Instead of detecting the pyrophosphate that is released during each nucleotide addition, these methods detect a change in pH that accompanies the reaction.

technology identifies the base that is added in 'real-time'. This is based on the colour emitted by the base while it resides within the active site of the immobilised DNA polymerase. The technology also enables detection of methylated bases, as the DNA polymerase reads them at a different rate when compared to the unmodified base, a change that can be quantified (see Chapter 6).

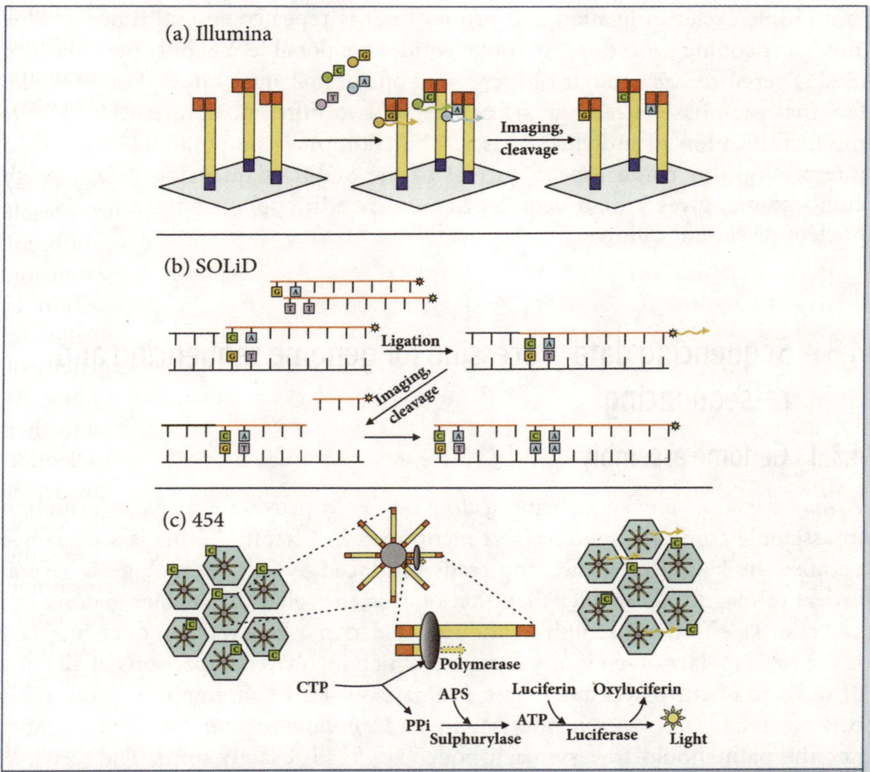

Figure 4.2 *Sequencing strategies for next-generation sequencers.* This figure shows a schematic representation of three commonly used sequencing strategies: (a) cyclic reversible termination with four fluorescent dyes, adopted by Illumina; (b) sequencing by ligation used by SOLiD sequencers; (c) pyrosequencing which underlies 454 sequencers. Single molecule real time sequencing is illustrated in Chapter 6. Figure credit: Avantika Lal, National Centre for Biological Sciences, India.

4. *Sequencing by ligation*: The sequencing by ligation procedure adopted by the SOLiD sequencing technology (Fig. 4.2 b) uses oligonucleotide probes in which the first two bases in their 3′ ends are read. These probes are added to the template–primer reaction mix, where the primer will be extended in its 5′ direction. The 5′-end of the primer is ligated to the 3′-end of the probe that is

hybridised adjacent to it. The remaining unhybridised probes are washed off. A fluorescent moiety attached to the 5′-end of the probe is cleaved off once the signal it emits is detected. This leaves the 5′-end free for another round of hybridisation and ligation. The procedure is repeated multiple times. Then the primer is offset by one base and the above procedure repeated, such that every base position is interrogated by two probes representing two different dinucleotides. The process of multiple cycles of ligation and primer reset is repeated several times. In this 'dibase' encoding procedure, 16 dinucleotides are possible. Yet only four dyes are used. Therefore, each colour can represent one of four nucleotides. However, the fact that each base is read in the context of two different dinucleotides allows the identification of individual bases. The additional computation involved in interpreting the 'colour–space' format of the SOLiD sequencing data, though cumbersome, gives a high degree of confidence in applications such as single nucleotide variant calling.

4.3 Sequencing data processing for genome sequencing and re-sequencing

4.3.1 Genome assembly

A major goal of any sequencing technology is to provide enough information to assemble complete genomes. As mentioned in Chapter 2, this is a challenge because the length of sequencing reads produced by any technology is several orders of magnitude smaller than that of even the relatively smaller genomes of bacteria. An effective assembly method applied to first-generation sequencing data is the overlap–layout–consensus method which involves comparisons of all pairs of reads to identify overlaps. These comparisons lead to a graph of connections between reads; the objective now is to find a *Hamiltonian path* through the graph, i.e., the path should traverse each node (i.e., read) exactly once. The fact that this requires an all-against-all pairwise comparison means that the computation required scales by the square of the number of reads used for the assembly. For example, a dataset with 10^3 reads will require 10^6 computations; one with 10^6 reads will require a whopping 10^{12} computations. Now, consider the fact that the commonly-used and cost-effective SOLiD and Illumina sequencers produce several hundred to thousand times more sequence reads for assembly than the first-generation Sanger sequencers–the computation required to assemble these reads using first-generation methods quickly appears formidable or even impractical. In addition, we do not know how to efficiently find a Hamiltonian path through a graph with millions of nodes. Further, the fact that the reads produced by these sequencers are 5–10-fold shorter than the Sanger reads means that overlaps between reads are relatively short. This raises the possibility of higher false positive rates during assembly. Finally, the error rates of these technologies are generally higher than Sanger sequencing,[6] and these need to be effectively corrected by the assembly algorithm.

An attractive solution that has become popular for assembling short reads into genomes is the use of what are called de Bruijn graphs. In the 1940s, a Dutch mathematician called Nicolaas de Bruijn became interested in finding the shortest circular string of characters that contains all possible substrings, each of length k, in a given alphabet (Fig. 4.3). The solution he came up with involved constructing a graph with all possible $(k-1)$-mers as nodes. Each k-mer was an edge directed from *nodeA* to *nodeB* if the $(k-1)$-mer in *nodeA* is a prefix, and that

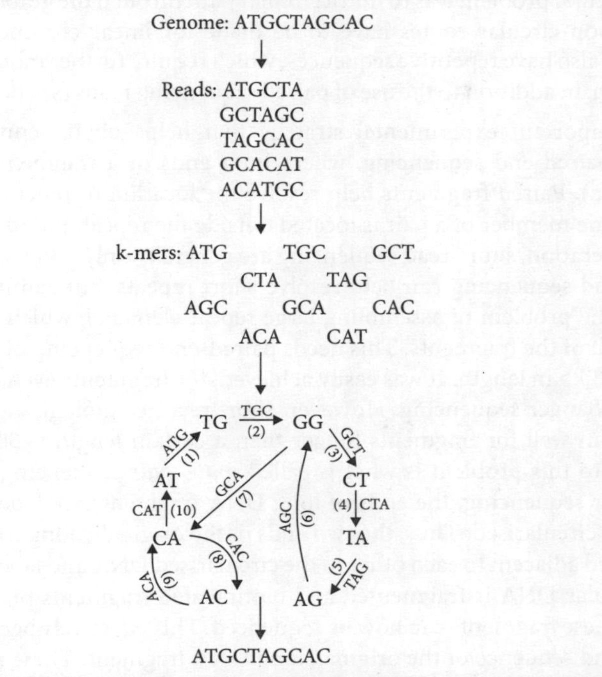

Figure 4.3 *de Bruijn graph for genome assembly.* This figure shows a toy example for the assembly of a sequence using de Bruijn graphs. In this example, a 10 nt genomic sequence is assembled from five reads, each of length 6. k-mers are identified and a de Bruijn graph with *(k–1)*-mers as nodes and k-mers as edges drawn as described in the text. A Eulerian path is traced through this network resulting in the reconstruction of the original genome sequence. Figure credit: Avantika Lal, National Centre for Biological Sciences, India.

6 However, they get better with every passing day.

in *nodeB*, a suffix of the *k*-mer. The answer to the stated problem now was to find a path through the graph that traverses each edge exactly once, in contrast to the Hamiltonian path which crosses each node exactly once. The great mathematician Leonard Euler had shown how to find such a path in a graph, nearly 300 years ago.[7] Assuming that every *k*-mer present in the genome has been sampled by the sequencing experiments, there exists a complete Eulerian path through the graph if each node, i.e., $(k - 1)$-mer, has equal numbers of incoming and outgoing edges.

Of course, real sequencing data face various issues that are not immediately dealt with by Euler's solution. For example, the sequencing is not complete and all possible *k*-mers are not sampled. Low coverage at certain loci and sequencing errors also lead to the genomic sequence being split into contigs. Though the original Euler problem was to find a circular path through the graph, modifications to find non-circular routes have to be made for linear chromosomes. Finally, genomes also have repetitive sequences, which require further modifications to the algorithm, in addition to the use of paired sequencing reads (see below), to resolve.

An important experimental strategy that helps obtain contiguous assemblies is paired-end sequencing, where both ends of a fragment are sequenced (Fig. 4.4 a). Paired fragments help resolve the location of repetitive elements, as long as one member of a pair is located outside the repeat. Because fragments for next-generation, short-read sequencing are generally only ~300 bp long, standard paired-end sequencing can help resolve short repeats, but cannot contribute to solving the problem of assembling large repeat elements, which are longer than the length of the fragments. This needs paired-end sequencing of long fragments, of say 2–3 kb in length. It was easily achieved, for fragments even longer than 2–3 kb, with Sanger sequencing. However, short read technologies are not designed to perform well for fragments longer than a certain length (~300–400 nt). The solution to this problem is what is called *mate–pair* sequencing (Fig. 4.4 b). In mate-pair sequencing, the ends of long DNA fragments are labelled with biotin and then circularised. Thus, the two ends of the originally long fragment are now positioned adjacent to each other in the circularised DNA and labelled with biotin. The circular DNA is fragmented and biotinylated fragments pulled down. Both ends of these fragments can now be sequenced. This effectively becomes a reversed paired-end sequence of the original, long DNA fragment. These mate–pairs help connect contings or scaffolds which are separated in the original assembly, and therefore contribute to genome closure.

---

[7] Euler had proposed this, in 1735, as a solution to what is known as the Bridges of Konigsburg problem. The old German city of Konigsburg had four land masses criss-crossed by rivers. These rivers could be crossed using seven bridges. Residents of this city wondered whether each land mass could be visited by traversing each of the seven bridges exactly once, finally returning to the starting point. Euler showed that this was not possible by using a graph representation wherein each land mass was a node, and each bridge an edge. This launched the now popular field of graph theory.

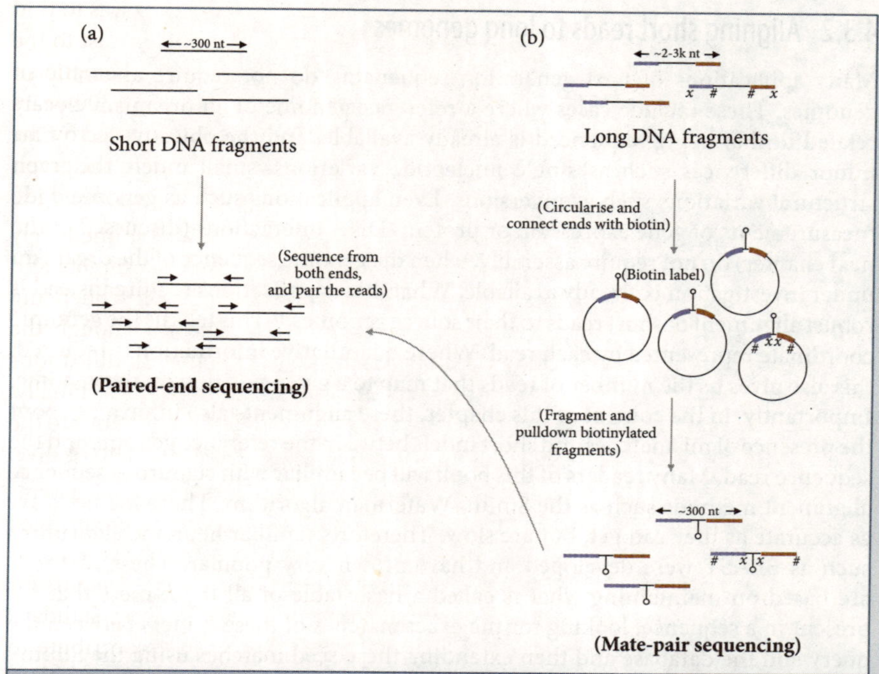

Figure 4.4 *Paired-end and mate-pair sequencing.* This figure shows a schematic of (a) paired-end and (b) mate-pair sequencing. Note that in mate-pair sequencing, circularization reverses the relative orientation of the two ends of the linear fragment. Whereas in paired-end sequencing, reads represent the two ends of the fragment outward-in, mate-paired reads represent the original long fragment inward-out. This change in orientation is marked by an *x* and a # in one of the fragments in (b).

Stable and robust software, which are free to download, use and modify, have been developed to assemble genomes from short reads using de Bruijn graphs. A particularly popular software called Velvet[8] appears to perform very well in assembling genomes, at least those of bacteria. The reader is referred to a review article for a concise description of de Bruijn graphs and their application to genome assembly.[9] In addition, the reader should also refer to a website[10] with practical tips for assembling bacterial genomes with Velvet.

8 www.ebi.ac.uk/~zerbino/velvet/; Zerbino and Birney. 2008. 'Velvet: Algorithms for de novo short read assembly using de Bruijn graphs.' *Genome Research* 18: 821–29.

9 Compeau P. E., Pevzner P. A. and Tesler G. 2011. 'How to apply de Bruijn graphs to genome assembly.' *Nature Biotechnology* 29: 987–91.

10 Nick Loman. *Tips for de novo bacterial genome assembly,* http://pathogenomics.bham.ac.uk/blog/2009/09/tips-for-*de-novo*-bacterial-genomeassembly/, accessed on 9th July, 2014.

4.3.2 Aligning short reads to long genomes

Many applications of next-generation sequencing do not require assembly of genomes. These include cases where a reference genome of an organism closely related to that being sequenced is already available, and the objective is to find minor differences such as single nucleotide variations, small indels and even structural variations such as inversions. Even applications such as genome-wide measurements of gene expression or protein–DNA interactions (discussed in the next chapter) do not require assembly, when the genome sequence of the organism under investigation is already available. What these applications require instead is robust alignment of short reads to their source genomes.[11] This tells us the genomic coordinate represented by each read. Where quantitative information is required, this also gives us the number of reads that map to a given position on the genome. Importantly, in the context of this chapter, these alignments also inform us about the presence of mismatches and short indels between the reference genome and the sequence read. Many readers of this book will be familiar with rigourous sequence alignment methods such as the Smith–Waterman algorithm. These methods are as accurate as they can get, but are slow. Therefore, familiar heuristic algorithms such as BLAST were developed and have grown very popular. These methods are based on maintaining what is called a hash table of all the *k*-mers that are present in a sequence, looking for the exact matches of these *k*-mers between the query and the database and then extending these seed matches using the Smith–Waterman algorithm. However, the BLAST algorithm was not primarily designed to align millions of short sequences to a long reference sequence. Therefore, it is not surprising that this algorithm is very time-inefficient in aligning short reads to a reference genome. This author, as a PhD student, attempted the exercise and it took well over a day for BLAST to align ~4 million 36-mer Illumina reads to the 4.6 Mb *E. coli* K12 genome. Though intelligent variants of the hash table-based algorithm of BLAST have been deployed for short read alignments (see software such as RMAP, MAQ and ZOOM), it has been suggested[12] that these are still too slow, especially in light of the fact that many projects involving next-generation sequencing produce data from a large number of samples (see for example, the 1,000 genomes, ENCODE and the modENCODE projects) and processing them using slow algorithms is not productive.

One approach that is widely used for short read alignments is based on a file compression algorithm called *Burrows–Wheeler transformation*, which is used in the poweful program Bzip in Unix/Linux environments (Fig. 4.5). Any given text *G*–for example the sequence of the reference genome–can be transformed

---

11 For a description of various algorithms used for such alignments, refer Li and Homer. 2010. 'A survey of sequence alignment algorithms for next-generation sequencing.' *Briefings in Bioinformatics* 2: 473–83.

12 Langmead B., Trapnell C., Pop M. and Salzberg S. L. 2009. 'Ultrafast and memory-efficient alignment of short DNA sequences to the human genome.' *Genome Biology* 10: R25.

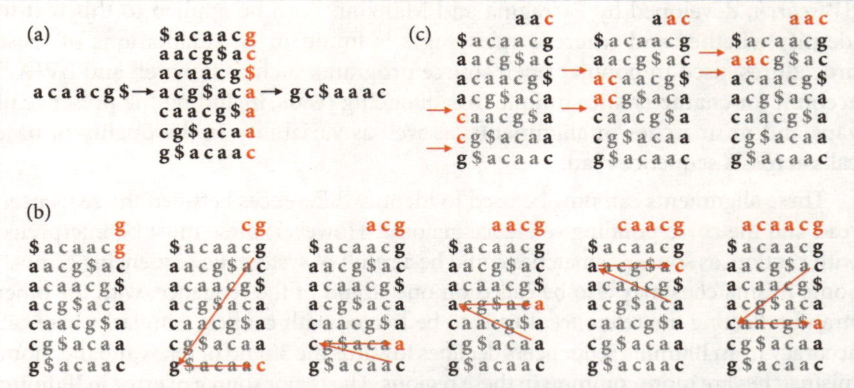

Figure 4.5 *Burrows–Wheeler transform for read mapping.* This figure shows a toy example of (a) constructing a Burrows-Wheeler transform of the sequence *acaacg*; (b) converting the Burrows-Wheeler transform back to the original sequence. See main text for further details; and (c) searching for a pattern *aac* within the sequence *acaacg*. This figure is reproduced under the Creative Commons Attribution license from Langmead, Trapnell, Pop and Salzberg. 2009. 'Ultrafast and memory-efficient alignment of short DNA sequences to the human genome.' *Genome Biology* 10: R25. © Langmead et al. 2009; licensee Biomed Central.

by the following procedure. The character $ is appended to *G* such that $ is not present in *G* and it is lexicographically smaller than all the characters in *G*. Then a matrix is built whose rows are the various cyclic rotations of the text *G*$. This matrix, which can be seen as a set of strings, is sorted in lexicographic order such that the first row begins with the character $. The Burrows–Wheeler transform of the text–referred to as BWT(G)–is the last column of the matrix. This is a compressible text, compression being possible by what is known as *run length encoding*. The 3 Gb human genome occupies only 2.2 Gb space on a hard disc, with only 1.3 Gb of memory required at runtime.[13] Information about the character composition of the text is sufficient to obtain the first column. Knowing only the first and the last column is sufficient to reconstruct the starting string *G*, as well as to find the occurrence and the position of a given substring *s*–for example, the sequence of a short read–in *G*. These are made possible by what is called the LF (last first) mapping property of the Burrows–Wheeler matrix. To quote Langmead and colleagues[14] in defining the LF mapping property, "The *i*th occurrence of character *x* in the last column corresponds to the same text character as the *i*th occurrence of *x* in the first column". A fast algorithm called

[13] These numbers are as reported by Langmead and colleagues for the program Bowtie, which uses the Burrows Wheeler Transform for alignments.

[14] Langmead B., Trapnell C., Pop M. and Salzberg S. L. 2009. 'Ultrafast and memory-efficient alignment of short DNA sequences to the human genome.' *Genome Biology* 10: R25.

BWsearch, developed by Ferragina and Manzini,[15] can be applied to this text to identify whether and where a substring *s* is found in *G*. Adaptations of these procedures, used in popular open source programs such as Bowtie[16] and BWA,[17] account for characteristics unique to sequencing reads, including the presence of gaps and mismatches in alignments, as well as variability in the quality of base calls across a sequence read.

These alignments can now be used to identify differences between the sequenced read and the corresponding reference genome. However, these must be interpreted with caution, as some mismatches could be a result of systematic sequencing errors.[18] Some mismatches may also be called on one strand of the sequence, with the other strand reflecting the reference; this is to be treated with caution. Similarly, base call accuracy from Illumina sequencers declines towards the 3′-end of reads and therefore, mismatches are more common in these regions. The major source of error in Illumina sequencing is what is known as *phasing* or *dephasing*, wherein base incorporation may fail at a few templates in any given cycle. Once this happens, the resulting 'frameshifting' is carried over to the rest of the sequencing run on that template sequence. As the number of cycles increases, the probability of finding such dephased templates in any given cluster increases and thus the chance of unambiguously identifying a single colour from that cluster decreases, resulting in lower base quality. Finally, an enrichment of errors in the context of specific sequence motifs have also been described. A recent high-profile work[19] identifying a large number of RNA editing events in human cells, based on mismatches between the genomic DNA and sequencing reads from the cDNA, has become controversial as the authors seem to have inadequately controlled for these factors. Therefore, it is important to be aware of the error characteristics of the sequencing technology and the properties of the alignment software being used.

---

15 http://people.unipmn.it/manzini/papers/focs00.html. Ferragina P. and Mancini G. 'Opportunistic data structures with applications.'

16 http://bowtie-bio.sourceforge.net/index.shtml; Langmead B., Trapnell C., Pop M. and Salzberg S. L., previously referred to in this chapter.

17 http://bio-bwa.sourceforge.net/; Li and Durbin. 2009. 'Fast and accurate short read alignment with Burrows-Wheeler transform.' *Bioinformatics* 25: 1754–1760.

18 (a) Nakamura K., Oshima T., Morimoto T., Ikeda S., Yoshikawa H., Shiwa Y., Ishikawa S., Linak M. C., Hirai A., Takahashi H., Altaf-Ul-Amin M., Ogasawara N. and Kanaya S. 2011. 'Sequence-specific error profile of Illumina sequencers.' *Nucleic Acids Research* 39: e90; (b) Meacham F., Boffelli D., Dhahbi J., Martin D. I., Singer M. and Pachter L. 2011. 'Identification and correction of systematic error in high-throughput sequence data.' *BMC Bioinformatics* 12: 451.

19 The original paper is Li M., Wang I. X., Li Y., Bruzel A., Richards A. L., Toung J. M. and Cheung V. G. 2011. 'Widespread RNA and DNA sequence differences in the human transcriptome.' *Science* 333: 53–58. Three technical comments, published in the same journal Science (Vol. 335, p. 1302), have contested the finding. These are by Lin W., Piskol R., Tan M. H., Li J. B., Pickrell J. K., Gilad Y., Pritchard J. K., Kleinman and Majewski. The authors of the original work have filed a response to these criticisms.

4.4 Case studies

With the economic viability and the relative ease of performing next-generation sequencing experiments, a humongous number of draft and complete genomes of microbes has become available in public databases. This has in fact prompted journals such as the *Journal of Bacteriology* (in collaboration with other journals from the American Society for Microbiology) to spawn a whole new journal called *Genome Announcements*, publishing 1–2-page reports on newly-sequenced genomes. The ambitious Human Microbiome Project, which provides high-quality draft to complete genome sequences of several thousand microbes that live on the human body, is largely run on data from 454 and Illumina sequencers. It is impossible to cite even a representative subset of these studies, and we will discuss a few pertinient studies that this author found instructive. Standard applications of next-generation sequencing in bacterial genome sequencing and re-sequencing are shown in Fig. 4.6.

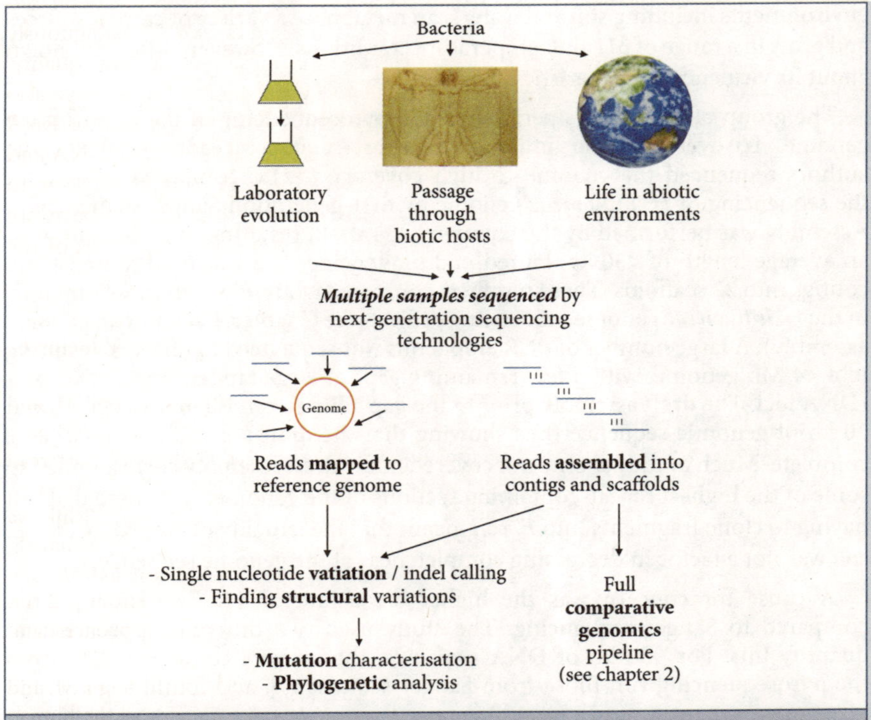

Figure 4.6 *Standard applications of next-generation sequencing.* This figure shows a schematic pipeline with various standard applications of next-generation sequencing to bacterial genome sequencing and re-sequencing.

4.4.1 Pyrosequencing-enabled complete genome sequence of *Acinetobacter baumanii*

In their early years, next-generation sequencers produced reads that were much smaller than the numbers quoted earlier in this chapter. Even 454 sequencers produced only ~100-mer reads, compared to the >400-mers that this technology provides today. In fact, today, even the shorter-read technology of Illumina sequencers routinely produces 100-mer reads, compared to the ~35-mers that it could provide in its early years. Therefore, next-generation sequencing technologies were confined to re-sequencing applications, with *de-novo* sequencing being a distant dream.

In 2007, Michael Snyder and colleagues from Yale University and Harvard Medical School published a report on the complete genome sequence of *Acinetobacter baumanii.*[20] *A. baumanii* is an opportunistic pathogen, which attracted particular American interest at that time because of widespread infection among their soldiers in Iraq. This gram negative bacterium lives in multiple environments including soil and water, can metabolise a variety of carbon sources and grow in a range of pH and temperature conditions. However, little was known about its virulence characteristics.

The group of researchers performed 454 pyrosequencing of the *A. baumanii* genome. To overcome the limitation of short sequence reads (~100 nt), the authors sequenced the genome to high coverage (>21x, compared to ~6x for the sequencing of *H. influenzae* genome by first-generation Sanger sequencing). Assembly was performed by the company 454 itself, resulting in <140 contigs of an average length of <30 kb. Paired-end sequencing data was used to link these contigs into 22 scaffolds. These numbers compare favourably with those obtained in the *H. influenzae* genome sequencing project (see Chapter 2, section on genome assembly). A large number of PCR amplicons was sequenced to fill gaps, resulting in a <4 Mb genome, with a few remaining gaps at large tandem repeats such as rDNA loci. The draft assembly prior to the gap-filling exercise missed only about 30 kb of genomic sequence thus showing that the draft assembly was ~99.24% complete. Much of the genome not covered in the draft assembly corresponded to some of the highest repeat-containing sections of the genome. Note here that not having to clone fragments into *E. coli* meant that the lethality of cloned loci for *E. coli* was not a factor in decreasing completeness of the genome sequence.

A cause for concern was the higher error rate of 454 sequencing when compared to Sanger sequencing. The study used two different approaches to quantify this. For ~30 kb of DNA sequence, the authors compared data from the pyrosequencing run those from Sanger sequencing, and found less than 30

[20] Smith M. G., Gianoulis T. A., Pukatzki S., Mekalanos J. J., Ornston L. N., Gerstein M. and Snyder M. 2007. 'New insights into Acinetobacter baumanii pathogenesis revealed by high-density pyrosequencing and transposon mutagenesis.' *Genes and Development* 21: 601–14.

mismatches, concluding that the sequencing was >99.92% accurate. A particular source of error in pyrosequencing is erroneous base calls in homopolymeric tracts, which would emerge from the addition of multiple bases of the same type to a homopolymeric template region during a single sequencing cycle. This would result in frameshifted genes during annotation. Such split genes could be identified by searching for tandem genes in the annotated genome which align to adjacent segments of the same gene in publicly available sequence databases. In this genome, 30 such instances were discovered.

The study identified ~30 horizontally acquired islands in the *A. baumanii* genome. One fourth of these carried putative antibiotic resistance determinants. However, the authors also note that the strain used for the sequencing was resistant only to β-lactams and state that the large number of putative resistance determinants was surprising. However, the genetic organisation of these resistance determinants seemed to differ from that in a multi-resistant clinical isolate, in which ~90% of the antibiotic resistance genes were clustered in a single locus. The authors also performed transposon mutagenesis to identify other genes involved in virulence.

4.4.2 On the track of pandemics: The genome of the aetiological agent of Black Death

Yersinia pestis is the causative agent of plague. It is primarily a pathogen of rodents, who act as reservoirs–it is transmitted by fleas. *Y. pestis* is a rod-shaped facultative anaerobe under *enterobacteriaceae.* Its genome, based on the numbers for two strains sequenced by Sanger sequencing, is ~4.6 Mb in size. The anaerobe is a descendent of *Yersinia pseudotuberculosis,*[21] which lacks two *Y. pestis*-specific virulence plasmids and rarely causes fatal diseases.

Though human disease due to *Y. pestis* is rare today and largely limited to certain endemic areas where there is high chance of human contact with infected rodents and fleas,[22] there have been three pandemics in the past. The first, called the *Justinian Plague*, happened between the 5th and the 8th centuries AD and

[21] Achtman M., Zurth K., Morelli G., Torrea G., Guiyoule A. and Carniel E. 1999. '*Yersinia pestis*, the cause of plague, is a recently emerged clone of Yersinia pseudotuberculosis.' *Proceedings of the National Academy of Sciences USA* 96: 14043–14048.

[22] An example is the plague epidemic that hit Surat in the Indian state of Gujarat in 1994. Caused by a heavy monsoon, which flooded gutters and left a large number of animal carcasses in its aftermath, the plague claimed >50 lives and caused widespread panic with hundreds of thousands of people migrating out of the city. Though *Y. pestis* could not be cultured by the Indian authorities at that time, it is thought that the disease was an *Yersinia*-dependent plague based on positive blood tests for the bacterium. However, this remains controversial, with several non-believers stating that there is much evidence that the disease was not plague.

spread from Egypt to other Mediterranean countries causing much loss to the Byzantine empire. The second pandemic, noted for the famous *Black Death* of 14th century Europe, was particularly severe and claimed 30–50% of the Europe population in a short span of five years. The third pandemic, which happened over a sixty-year period from the end of the 19th century, originated in Hong Kong, and also led to the identification of *Y. pestis* by Koch-ian methods.

An important question in plague research is the relationship of the currently circulating strains of *Y. pestis* to those associated with the historic pandemics. Bos and co-workers[23] addressed this question by obtaining draft genome sequences of *Y. pestis* from dental remains of victims of the 14th century plague. An important source of errors in sequencing ancient DNA is the spontaneous process of deamination, which converts cytosine to uracil. Bos and colleagues accounted for this by treating extracted genomic DNA with uracil-DNA-glycosylase, which eliminates uracil from DNA molecules, as well as endonuclease VIII, which cleaves at damaged pyrimidines. To enrich for DNA from *Y. pestis*, the authors used an array capture technique. Here, oligonucleotide probes representing the genome of the extant *Y. pestis* CO92 were used to purify complementary DNA from the ancient DNA sample. The resulting DNA was subjected to paired-end Illumina sequencing for 76 cycles. Since the DNA was heavily fragmented with an average length of only 55 bp, overlaps between the sequences from the two ends were used as a quality filter, and the read pairs were fused together. Finally, ~2.3 million high-quality chromosomal reads were used for further analysis. The reads were processed using (a) mapping of these fused reads to the reference genome of *Y. pestis* CO92 with the Burrows–Wheeler aligner BWA, and (b) reference-assisted assembly of reads using the program Velvet. Using these methods, the authors were able to assemble a draft genome of ~4.0 Mb. As a result of the capture procedure used, DNA segments that were unique to the ancient DNA could not be recovered. From these mapping and alignment data, the authors could detect single nucleotide variations between the ancient *Y. pestis* and the extant reference genome (Fig. 4.6). They could also describe larger structural variations, where different segments of DNA from the same contig in the ancient *Y. pestis* genome aligned against chromosomally distant portions of the reference genome.

Bos and co-workers found only 97 single nucleotide variations between the ancient genome and that of the reference *Y. pestis* CO92, a surprisingly small number. Remarkably at all these 97 variant sites, the nucleotide in the ancient genome corresponded exactly to that in *Y. pseudotuberculosis*, the ancestor of *Y. pestis*. However, the fact that the agent of Black Death was in fact *Y. pestis* and not *Y. pseudotuberculosis* could be confirmed from the presence of the two *Y. pestis*-

---

[23] Bos K. I., Schuenemann V. J., Golding G. B., Burbano H. A., Waglechner N., Coombes B. K., McPhee J. B., DeWitte S. N., Meyer M., Schmedes S., Wood J., Earn D. J., Herring D. A., Bauer P., Poinar H. N. and Krause J. 2011. 'A draft genome of Yersinia pestis from victims of Black Death.' *Nature* 478: 506–10.

specific plasmids. Phylogenetic analysis of the ancient *Y. pestis*, in the context of the genomes of 17 extant *Y. pestis* strains and the ancestral *Y. pseudotuberculosis*, placed the agent of Black Death only *two substitutions from the root of all extant human pathogenic Y. pestis* (Fig. 4.7). However, *Y. pestis* isolated from one of the four Black Death victims carried three additional derived nucleotides seen in extant isolates, suggesting microevolution of *Y. pestis* within the context of the 14th century outbreak. Here we note that another study,[24] based on Sanger and pyrosequencing of several isolates tracked the origin of extant *Y. pestis* to

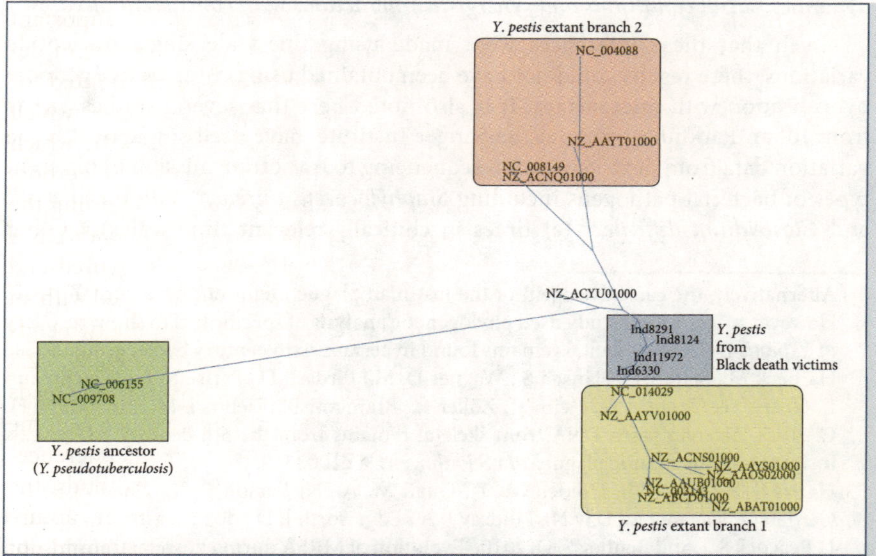

Figure 4.7 *A phylogenetic tree of Yersinia pestis*. This simple phylogenetic tree was constructed using ClustalW Alignment and Phylogeny on the European Bioinformatics Institute web server, and rendered using the Trex Newick Viewer online (http://www.trex.uqam.ca/index.php?action=newick). Data on polymorphic sites, and the definition of the two branches of currently circulating *Y. pestis* strains, were derived from the deep-sequencing-based work by Bos and co-workers (Bos et al. 2011). *Y. tuberculosis* (ancestors of *Y. pestis;* green), *Y. pestis* from Black Death victims (blue), and the two branches of extant *Y. pestis* pathogens (red and yellow) are marked.

[24] Morelli G., Song Y., Mazzoni C. J., Eppinger M., Roumagnac P., Wagner D. M., Feldkamp M., Kusecek B., Vogler A. J., Li Y., Cui Y., Thomson N. R., Jombart T., Leblois R., Lichtner P., Rahalison L., Petersen J. M., Balloux F., Keim P., Wirth T., Ravel J., Yang R., Carniel E. and Achtman M. 2010. 'Phylogenetic diversity and historical patterns of pandemic spread of Yersinia pestis.' *Nature Genetics* 42: 1140–1143.

China, to a period close to that of Black Death. Finally, assigning divergence times to the phylogenetic tree placed the ancestral *Y. pestis* sequence to a period from the end of the 13th century to mid-fourteenth century. This suggested that the causative agent of the much older Justinian plague was probably very divergent from the ancestor of the currently circulating *Y. pestis* strains.[25]

Though human plague is relatively rare today, the similarity of currently-circulating strains to those of the agent of the pandemic plague of the 14th century is interesting and allowed the authors to reiterate the role of a variety of factors to infectious disease including *genetics of the host population, climate, vector dynamics, social conditions and synergistic interactions with concurrent diseases.*

Given that these inferences were made using fine-scale single nucleotide variations, these results could not have been obtained using comparative genome hybridisation with microarrays. It is also noted here that several studies, many from Julian Parkhill's group at the Sanger Institute, have used single nucleotide variation data from next-generation sequencing to track transmission of different types of bacterial pathogens including *Staphylococcus aureus,*[26] *Vibrio cholerae*[27] and *Clostridium difficile,*[28] (at times in clinically-relevant time-scales).[29] Other

---

25 Alternatively, the causative agent of the Justinian plague might not have been *Y. pestis.* However, a very recent study used phylogenetic analysis of specific loci to show recovery of *Y. pestis* DNA from dental remains found in certain sixth century burial grounds. See Harbeck M., Seifert L., Hänsch S., Wagner D. M., Birdsell D., Parise K. L., Wiechmann I., Grupe G., Thomas A., Keim P., Zöller L., Bramanti B., Riehm J. M. and Scholz H. C. 2013. '*Yersinia pestis* DNA from skeletal remains from the 6th century AD reveals insights into Justinianic plague.' *PLoS Pathogens* 9: e1003349.

26 Harris S. R., Feil E. J., Holden M. T., Quail M. A., Nickerson E. K., Chantratita N., Gardete S., Tavares A., Day N., Lindsay J. A., Edgeworth J. D., de Lencastre H., Parkhill J., Peacock S. J. and Bentley S. D. 2010. 'Evolution of MRSA during hospital transmission and intercontinental spread.' *Science* 327: 469–74.

27 Mutreja A., Kim D. W., Thomson N. R., Connor T. R., Lee J. H., Kariuki S., Croucher N. J., Choi S. Y., Harris S. R., Lebens M., Niyogi S. K., Kim E. J., Ramamurthy T., Chun J., Wood J. L., Clemens J. D., Czerkinsky C., Nair G. B., Holmgren J., Parkhill J. and Dougan G. 2011. 'Evidence for several waves of global transmission in the seventh cholera pandemic.' *Nature* 477: 462–77.

28 He M., Sebaihia M., Lawley T. D., Stabler R. A., Dawson L. F., Martin M. J., Holt K. E., Seth-Smith H. M., Quail M. A., Rance R., Brooks K., Churcher C., Harris D., Bentley S. D., Burrows C., Clark L., Corton C., Murray V., Rose G., Thurston S., van Tonder A., Walker D., Wren B. W., Dougan G. and Parkhill J. 2010. 'Evolutionary dynamics of *Clostridium difficile* over short and long time scales.' *Proceedings of the National Academy of Sciences USA*. 107: 7527–7532.

29 Köser C. U., Holden M. T., Ellington M. J., Cartwright E. J., Brown N. M., Ogilvy-Stuart A. L., Hsu L. Y., Chewapreecha C., Croucher N. J., Harris S. R., Sanders M., Enright M. C., Dougan G., Bentley S. D., Parkhill J., Fraser L. J., Betley J. R., Schulz-Trieglaff O. B., Smith G. P. and Peacock S. J. 'Rapid whole-genome sequencing for investigation of a neonatal MRSA outbreak.' *New England Journal of Medicine* 366: 2267–2275.

examples of deep-sequencing enabled rapid genomic characterisation of outbreak strains include those of *V. cholerae* responsible for an outbreak in Haiti in 2010,[30] and of *E. coli* which caused food-poisoning in Europe in 2011[31] (Fig. 4.8).

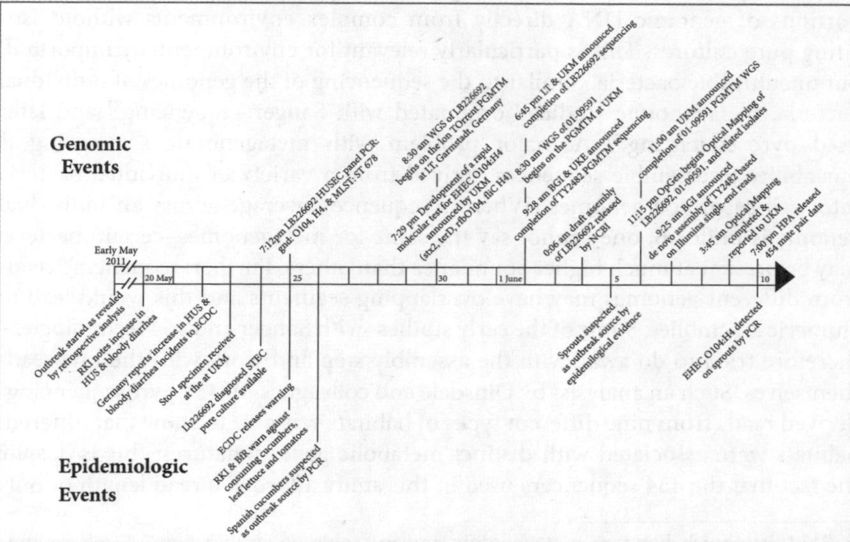

Figure 4.8 *Rapid genomic characterisation of an E. coli outbreak.* This figure shows the timeline of events leading from the recognition of an E. coli outbreak to the announcement of the release of the causative bacterium's draft genome within around a month. The sequencing – on an Ion Torrent benchtop sequencer – to the assembly took only a couple of days. This figure is reproduced under the Creative Commons Attribution license from Mellmann A., Harmsen D., Cummings C. A., Zentz E. B., Leopold S. R., et al. 2011. 'Prospective genomic characterization of the German enterohemorrhagic *Escherichia coli O104: H4* outbreak by rapid next generation sequencing technology.' *PLoS One* 6(7): e22751. © Mellman et al. 2011.

30 Katz L. S., Petkau A., Beaulaurier J., Tyler S., Antonova E. S., Turnsek M. A., Guo Y., Wang S., Paxinos E. E., Orata F., Gladney L. M., Stroika S., Folster J. P., Rowe L., Freeman M. M., Knox N., Frace M., Boncy J., Graham M., Hammer B. K., Boucher Y., Bashir A., Hanage W. P., Van Domselaar G. and Tarr C. L. 2013. 'Evolutionary dynamics of Vibrio cholerae O1 following a single-source introduction to Haiti.' *mBio* 4: e00398–13.

31 Mellmann A., Harmsen D., Cummings C. A., Zentz E. B., Leopold S. R., Rico A., Prior K., Szczepanowski R., Ji Y., Zhang W., McLaughlin S. F., Henkhaus J. K., Leopold B., Bielaszewska M., Prager R., Brzoska P. M., Moore R. L., Guenther S., Rothberg J. M. and Karch H. 2011. 'Prospective genomic characterization of the German enterohaemorrhagic *Escherichia coli* O104: H4 outbreak by rapid next generation sequencing technology.' *PLoS One* 6: e22751.

4.4.3 From community genomes to complete genomes to single-cell genomes

A relatively recent development in genomics is our ability to sequence large portions of genomic DNA directly from complex environments without isolating pure cultures. This is particularly relevant for environmentally important, but uncultivable bacteria. Similar to the sequencing of the genomes of individual bacteria, metagenomic studies[32] originated with Sanger sequencing[33] and later used pyrosequencing.[34] A major problem with metagenomic sequencing is our ability to assemble sequences derived from a variety of individual bacteria into well-separated genomes. Whereas sequence coverage across an individual genome is uniform, one cannot say the same for metagenomes–certain bacteria may be present at much higher abundance than others. Further, sequencing reads from different genomes may have overlapping segments and this would lead to chimeric assemblies. Some of the early studies–with Sanger and 454 technologies–therefore tried to do away with the assembly step and work with the raw reads themselves. Such an analysis–by Dinsdale and colleagues–of 454 pyrosequencing-derived reads from nine different types of habitats was able to show that different habitats were associated with distinct metabolic gene signatures; this is despite the fact that the 454 sequencers used in this study afforded a read length of only

---

32 We distinguish between metagenomic and microbiome studies here. A microbiome study need not attempt whole genome sequencing of the genetic content of bacterial communities. Instead, these studies perform sequencing of PCR amplicons–typically from the 16S rDNA–to identify and quantify the types of bacteria that are present in the sample. For example, Jeffrey Gordon's group used pyrosequencing of a 16S rRNA amplicon to catalogue the bacterial diversity of different anatomical parts of the human body over a period of time (Costello E. K., Lauber C. L., Hamady M., Fierer N., Gordon J. I. and Knight R. 2009. 'Bacterial community variation in human body habitats across space and time.' *Science* 326: 1694–1697). The discussion in this book will emphasise metagenomic studies which, on the other hand, also provide a catalogue of most genes present in a bacterial community by sequencing the entire pool of genomic DNA that can be isolated from a sample. A hybrid approach involves microbiome sequencing, coupled with whole-genome sequencing of culturable reference isolates (see data from the Human Microbiome Project at http://hmpdacc.org/), or of single bacterial cells isolated from a community (described below).

33 See for example, Tringe S. G., von Mering C., Kobayashi A., Salamov A. A., Chen K., Chang H. W., Podar M., Short J. M., Mathur E. J., Detter J. C., Bork P., Hugenholtz P. and Rubin E. M. 2005. 'Comparative metagenomics of microbial communities.' *Science* 308: 554–57.

34 Dinsdale E. A., Edwards R. A., Hall D., Angly F., Breitbart M., Brulc J. M., Furlan M., Desnues C., Haynes M., Li L., McDaniel L., Moran M. A., Nelson K. E., Nilsson C., Olson R., Paul J., Brito B. R., Ruan Y., Swan B. K., Stevens R., Valentine D. L., Thurber R. V., Wegley L., White B. A. and Rohwer F. 2008. 'Functional metagenomic profiling of nine biomes.' *Nature* 452: 629–32.

~100. We do not discuss these here. Instead, we focus on recent developments by which we can obtain draft to complete genome sequences of individual members of a bacterial community, a field which has been enabled by next-generation sequencing and by single-cell genomics.

The complexity of environmental communities at times required high sequence depth to sample enough members of such communities. Therefore, it was not long before short read deep-sequencing technologies such as Illumina entered the field in a big way. A pioneering example is the sequencing of the gut microbiomes of over a hundred individuals using high-coverage Illumina sequencing.[35] However, more than the above-mentioned examples with Sanger sequencing, shorter reads do benefit immensely from genome assembly, and as a result, much research has gone into algorithms for assembling individual genomes from metagenomes. The literature discussed below will emphasise this recent development in the large and fast-growing body of metagenomic work. A schematic representation of these approaches is shown in Fig. 4.9.

Hess and colleagues[36] from Eddy Rubin's group generated genomic DNA libraries from the microbiota present in the rumen of cows. The primary objective of this work was to genetically characterise a bacterial community capable of degrading cellulosic plant material. Using Illumina sequencers producing read lengths in the 100–125-mer range, the authors generated over 260 Gb of sequence data. After filtering sequencing reads to remove those containing adapter sequences, low complexity regions, and tandem repeats, the authors further discarded reads containing low occurrence of 31-mers, to reduce the adverse effects of possible sequencing errors and low abundance species, which will anyway be refractory to assembly. Finally, ~110 Gb of sequence data were assembled using the de Bruin graph-based software Velvet to obtain nearly 180,000 scaffolds, each of length greater than 1 kb. The assembly amounted to nearly 2 Gb of genomic DNA sequence. Scaffolds that were believed to be chimeras from different genomes were split at specific breakpoints, based on–for example–differential sequencing coverage on either side of the breakpoint. Finally, a set of ~26,000 scaffolds, each of length greater than 10 kb, was

[35] Qin J., Li R., Raes J., Arumugam M., Burgdorf K. S., Manichanh C., Nielsen T., Pons N., Levenez F., Yamada T., Mende D. R., Li J., Xu J., Li S., Li D., Cao J., Wang B., Liang H., Zheng H., Xie Y., Tap J., Lepage P., Bertalan M., Batto J. M., Hansen T., Le Paslier D., Linneberg A., Nielsen H. B., Pelletier E., Renault P., Sicheritz-Ponten T., Turner K., Zhu H., Yu C., Li S., Jian M., Zhou Y., Li Y., Zhang X., Li S., Qin N., Yang H., Wang J., Brunak S., Doré J., Guarner F., Kristiansen K., Pedersen O., Parkhill J., Weissenbach J., MetaHIT Consortium, Bork P., Ehrlich S. D. and Wang J. 2010. 'A human gut microbial gene catalogue established by metagenomic sequencing.' *Nature* 464: 59–65.

[36] Hess M., Sczyrba A., Egan R., Kim T. W., Chokhawala H., Schroth G., Luo S., Clark D. S., Chen F., Zhang T., Mackie R. I., Pennacchio L. A., Tringe S. G., Visel A., Woyke T., Wang Z. and Rubin E. M. 2011. 'Metagenomic discovery of biomass-degrading genes and genomes from cow rumen.' *Science* 331: 463–67.

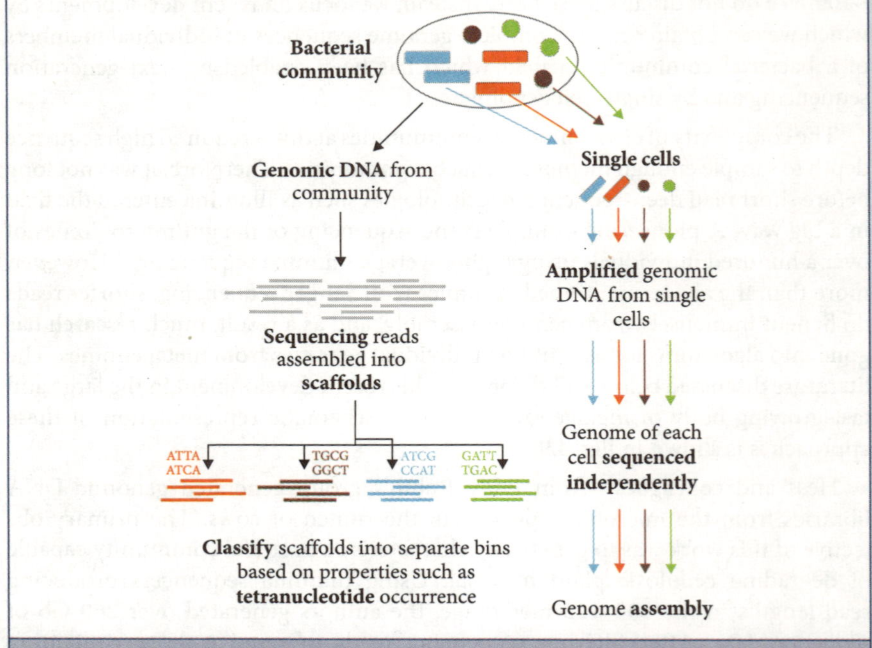

Figure 4.9 *From communities to single-cell genomes.* In one approach to culture-free genome sequencing, genomic DNA is isolated directly from complex bacterial communities and sequenced. The challenge is to separate out reads belonging to one genome from those from other genomes. This is typically achieved using tetranucleotide occurrence, which has been shown to be a sensitive discriminator of distinct genomes. Given sufficient sequencing coverage, it has become possible to obtain draft to complete genome sequences of a single bacterial species from such complex meta-genomes. In the second approach, single bacterial cells are isolated from communities, genomic DNA from each single cell amplified and then sequenced.

used to assemble draft genomes of specific components of this community. A major bottleneck in metagenomic studies is the difficulty in classifying sequences based on their source organism. However, a study performed in 2004[37] showed that the occurrence of tetra-nucleotides could discriminate between genomes. Hess and colleagues calculated the occurrence of tetranucleotides in the scaffolds from their assembly. They built a tetranucleotide frequency matrix–for example, each row being a scaffold and each column, a tertranucleotide, with each cell showing the frequency of a tetranucleotide in a scaffold. This matrix was clustered and groups of scaffolds, presumably from the same or similar species,

[37] Teeling H., Meyerdierks A., Bauer M., Amann R. and Glöckner F. O. 2004. 'Application of tetranucleotide frequencies for the assignment of genomic fragments.' *Environmental Microbiology* 6: 938–47.

binned together. The authors mention that the use of tetranucleotides for genome binning cannot separate closely-related organisms. However, on the assumption that the number of reads that map to a scaffold (read coverage) is a measure of the abundance of the source organism in the sample, the authors split tetranucleotide frequency-based bins with bimodal distribution of read coverage into separate bins. This is justified by the common sense argument that two regions from the same chromosome cannot have different abundance levels in the same sample– which should be largely true with some exceptions caused by biases during PCR amplification, or rapidly-dividing organisms in the exponential phase where the origin of replication fires multiple times in the same cell division cycle. Finally, the set of scaffolds corresponding to each bin was defined as a draft assembly for a particular genome. To assess the completeness of their draft assemblies, the authors matched each assembly to the phylogenetic order to which the source organism belongs. Then, they downloaded the complete genome sequences of all members of the order from public databases and estimated the core genome defined by the set of genes present in all members of the order. Next, they calculated the proportion of these core genes that could be recovered from their draft assemblies. This calculation revealed that their top 15 draft assemblies were between 60% and 93% complete. In a more recent study, Iverson and coworkers[38] assembled 14 draft genomes of bacteria and archaea from a marine environment, using ~60 Gb of 50-mer data from a SOLiD sequencer. The assemblies were built using a tetranucleotide frequency-based binning method. Taking this a step further, the authors fully closed the genome of an uncultured archaea using this assembly as a starting point.

At this point, the readers are directed to approaches for sequencing the genomes of single bacterial cells from complex communities. In this approach, individual cells are separated by flow cytometry into 384-well plates. The DNA from single cells is amplified using what is known as multiple displacement amplification. In this approach, amplification of genomic DNA is performed using random hexamers and a high fidelity Φ29 DNA polymerase. When the extending DNA polymerase encounters a replication starting site (another primer), it displaces the newly-synthesised strand, which can now act as a template for further amplification. The process results in a highly branched DNA network, which can be resolved by S1 nuclease. The procedure is capable of producing 1–2 ug of DNA from a single cell. This amplified DNA product can then be used for making libraries for genome sequencing. Recently, a group led by Roger Lasken and Craig Venter *pyro*-sequenced the genome of a pathogen *Porphorymonas gingivalis* isolated from a biofilm using this approach.[39]

---

38 Iverson V., Morris R. M., Frazar C. D., Berthiaume C. T., Morales R. L. and Armbrust E. V. 2012. 'Untangling genomes from metagenomes: Revealing an uncultured class of marine euryarchaeota.' *Science* 335: 587–90.

39 McLean J. S., Lombardo M. J., Ziegler M. G., Novotny M., Yee-Greenbaum J., Badger J. H., Tesler G., Nurk S., Lesin V., Brami D., Hall A. P., Edlund A., Allen L. Z., Durkin S., Reed S., Torriani F., Nealson K. H., Pevzner P. A., Friedman R., Venter J. C. and

4.4.4 Bacteria evolving in the laboratory

Bacteria have short doubling times and reach very high population densities in laboratory environments. These features allow them to be excellent models for playing evolutionary games in the laboratory. For example, can I grow a culture in the lab under a certain stress and eventually obtain a population that can grow efficiently under this stress. Or for that matter, can I grow a population of bacteria under standard laboratory conditions long enough for it to start growing faster than it normally does? If the answer is 'yes', can I now characterise mutations that make these possible (Fig. 4.6)?

A classical example of laboratory evolution experiments for bacteria is that done by Richard Lenski and colleagues. These researchers have been evolving *E. coli* in aerated batch culture in minimal media for over 50,000 generations now, with frozen stocks established every 500 generations. The group has discovered many novelties in the evolving population. One remarkable example is described here. The minimal medium in which the bacteria are grown also contains citrate, an intermediate of the TCA cycle. However, *E. coli* cannot metabolise citrate under aerobic conditions, primarily because of its inability to transport the molecule into the cell. This is because, a gene *citT*, coding for a citrate transporter, is not expressed under these conditions. The laboratory evolution experiment however generated a version, at ~31,500 generations, which could now metabolise citrate (Cit^+).[40] This phenotype became stronger over the next ~1,500 generations at which point the Cit^+ variant became dominant in the population. Because of the high levels of citrate in the medium, the development of the Cit^+ variant also led to a large increase in the population size of the culture. Using Illumina sequencers, the group sequenced the genomes of several populations of *E. coli* obtained at various frames from the evolution experiment.[41] Mapping of sequence reads to the reference genome enabled identification of small variations. Larger amplifications and deletions were identified by examining the read coverage of specific regions of the genome. The genome of an early Cit^+ strain showed the presence of a duplication in the locus containing the *citT* transporter gene. The duplication was such that it brought one copy of *citT* under the promoter of a gene called *rnk*, which is expressed under aerobic conditions. Therefore, amplication of a locus results in 'promoter capture', which enables the expression of an otherwise silent gene. Sequencing of other lines with Cit^+ phenotype all revealed promoter capture

---

Lasken R. S. 2013. 'Genome of the pathogen Porphyromonas gingivalis recovered from a biofilm in a hospital sink using a high-throughput single-cell genomics platform.' *Genome Research* 23: 867–77.

[40] Blount Z. D., Borland C. Z. and Lenski R. E. 2008. 'Historical contingency and the evolution of a key innovation in an experimental population of *Escherichia coli*.' *Proceedings of the National Academy of Sciences USA* 105: 7899–7906.

[41] Blount Z. D., Barrick J. E., Davidson C. J. and Lenski R. E. 2012. 'Genomic analysis of a key innovation in an experimental *Escherichia coli* population.' *Nature* 489: 513–18.

by the *citT* gene, but from different loci: (a) the insertion of an IS3 element 5′ of *citT*, with the IS3 element providing an outward promoter activity supporting expression of adjacent genes; (b) an inversion that places *citT* under the promoter for fimbrial genes; (c) a deletion of a gene called citG, which is upstream of *citT*, and which might create a new promoter. The authors also examined the processes that helped strengthen the Cit$^+$ phenotype after its establishment. Though the answer is not clear, an attractive mutation was seen in the gene *arcB*, encoding a histidine kinase, whose inactivation would lead to over expression of the TCA cycle, which might make citrate metabolism more efficient. The study also discusses epistatic phenomenon that enable the establishment of Cit$^+$ in the first place. Readers are directed to the original publication for further details.

Bacteria gain resistance to antibiotics in the laboratory via mutations. Next-generation sequencing allows us to characterise such mutations and also track the properties of evolutionary pathways leading to resistance. Roy Kishony's group at Harvard University has pursued such work. In one such study, Toprak and co-workers performed experiments producing resistance in *E. coli* to three antibiotics: Chloramphenicol (CHL), doxycycline (DOX) and trimethoprim (TMP).[42] The first two antibiotics target the ribosome, whereas the third targets the essential metabolic enzyme DHFR (dihydrofolate reductase). To perform the evolution experiment, the authors used a variant version of a continuous culture chemostat (termed morbidostat), in which the culture growth rate was maintained by changing antibiotic concentrations. As the population gains resistance to an antibiotic concentration *C*, it would grow better than the starting culture in *C*; but a feedback circuit that increases the antibiotic concentration appropriately would ensure that the growth rate is maintained at the pre-determined level. These evolution experiments first showed that resistance to CHL and DOX increased smoothly with time, whereas that for TMP increased in a step-wise fashion. Sequencing of the genomes of multiple populations using Illumina sequencers, followed by mapping of sequencing reads to the reference *E. coli* genome, showed that resistance to CHL and DOX emerged through multiple routes, with mutations in genes involved in membrane transport and those involved in transcription and translation. As a result of these generic mutations, clones resistant to CHL also showed resistance to DOX and vice-versa. Though these two antibiotics target the ribosome, no ribosomal mutation could be detected. This could be because of fitness costs that such mutations could cause. On the other hand, most mutations for TMP resistance were confined to the *dhfr* gene and its promoter. This explains the step-wise increase in resistance, where a population with the first mutation waits for a while to gain a second, rare mutation within the same gene. The promoter mutations were likely to increase expression of the *dhfr* gene, whereas

42 Toprak E., Veres A., Michel J. B., Chait R., Hartl D. L. and Kishony R. 2012. 'Evolutionary paths to antibiotic resistance under dynamically sustained drug selection.' *Nature Genetics* 44: 101–105.

the mutations in the gene body mapped close to the active site. The identity of the mutations in the *dhfr* gene could be reproduced in other replicate populations as well. Several of these mutations had also been previously seen in clinical isolates. These allowed the authors to propose that TMP resistance evolves by 'sequential fixation of mutations' through 'ordered pathways'.

4.4.5 Bacteria evolving in their biotic hosts

Evolutionary pathways and endpoints observed in controlled conditions in the laboratory may not necessarily reflect those that occur in natural habitats, including the human hosts of pathogens. The latter is not only important, but also difficult to study. However, next-generation sequencing, together with careful experimental designs, have enabled studies of bacterial evolution in their natural hosts at high temporal resolution (Fig. 4.6). Here we explore the outcomes of two such studies.

The first study we discuss was pursued once again by the Kishony group at Harvard (see previous section).[43] In this work, Leiberman and colleagues studied genetic adaptation of a single strain of *Burkholderia dolosa* that had caused a minor epidemic in Boston (USA) in the 1990s. What enabled this study was the fact that bacteria isolated from each of 39 infected patients were frozen and stored by a hospital. Lieberman and colleagues sequenced the genomes of 112 *B. dolosa* isolates from 14 different patients using an Illumina sequencer, obtaining 75-mer single-end sequence reads amounting to an average coverage of ~37x. Since the authors were interested in single nucleotide variations and not in larger structural variations and mobile genetic elements, the analysis was driven by the alignment of their sequence reads to a *B. dolosa* reference genome. A phylogenetic analysis of these isolates based on single nucleotide variations showed clustering of bacteria isolated from the same individual (Fig. 4.10). This allowed the authors to infer the last common ancestor of bacteria from each patient, and thus study the evolution of the bacteria within the same individual. The analysis showed that a pair of isolates obtained from the blood of the same individual originated from two distinct lung isolates. Further, phylogenetic relationships among the last common ancestors of each patient enabled the construction of a network of transmission among the subjects. The authors were also able to make correlations between specific phenotypes and genotypes. For example, the isolates were different from each other in their levels of resistance to ciprofloxacin. Comparison between resistant and sensitive genotypes mapped the causative mutation to two residues in DNA gyrase. The same mutations were observed in different subjects, with

[43] Lieberman T. D., Michel J. B., Aingaran M., Potter-Bynoe G., Roux D., Davis M. R. Jr., Skurnik D., Leiby N., LiPuma J. J., Goldberg J. B., McAdam A. J., Priebe G. P. and Kishony R. 2011. 'Parallel bacterial evolution within multiple patients identifies candidate pathogenicity genes.' *Nature Genetics* 43: 1275–1280.

phylogenetic analysis–in relation to the last common ancestor within each host–indicating that these were independently acquired within the host post-infection. Similarly, the authors found that the infecting parent strain did not produce the O-antigen, but acquired a mutation (in 9 hosts), which allowed it to express the antigen, a player in the virulence of the bacterium. Finally, the authors identified nearly 20 additional genes accumulating multiple mutations in several individuals; these genes might represent important pathways involved in pathogenesis.

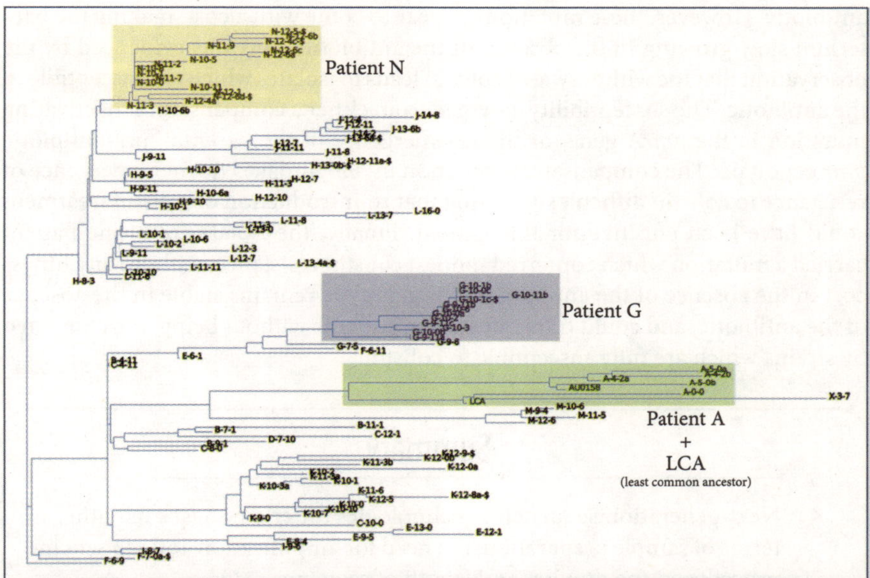

Figure 4.10 *Pathogen evolution in hosts.* A phylogenetic tree – similar to that in Fig. 4.7 – of *Burkhoderia dolosa* isolated from different patients from a minor epidemic in Boston. Note the clustering of isolates from the same individual (yellow and blue) and the result that the last common ancestor of these strains is closest to the isolates from patient A (green). These were drawn as described for Fig. 4.7, based on data published by Lieberman et al. 2011.

In a very recent study, researchers from the National Institutes of Health, USA, tracked the emergence of resistance to the peptide antibiotic colistin in patients infected with *Acinetobacter baumanii*. In this study, Snitkin and co-workers sequenced the genomes of *A. baumanii* isolated from four patients at different time-points during and after colistin treatment.[44] Sequencing reads obtained

44 Snitkin E. S., Zelazny A. M., Gupta J., NISC Comparative Sequencing Program, Palmore T. N., Murray P. R. and Segre J. A. 2013. 'Genomic insights into the fate of colistin resistance and *Acinetobacter baumanii* during patient treatment.' *Genome Research* 23: 1155–1162.

using a 454 pyrosequencer were assembled into contigs and then aligned against the genome of the initial isolate for that individual. From these alignments, single nucleotide variants and indels were obtained. Isolates showing resistance to the antibiotic carried several mutations all mapping to the gene *pmrB*, a sensor kinase which phosphorylates a downstream transcription factor encoded by *pmrA*. Further analysis showed that these mutations make PmrB constitutively active. PmrA activates transcription of *pmrC*, which encodes a protein that modifies the LPS thus decreasing the electrostatic attraction between the cell surface and the antibiotic. However, these mutations appear to come with a cost, making the bacterium slow growing in the absence of the antibiotic. This was evidenced by the observation that the withdrawal of colistin leads to isolates which are susceptible to the antibiotic. The susceptibility emerged from either a compensatory inactivating mutation in the *pmrA* gene, or the re-emergence of the parental, pre-antibiotic *pmr* genotype. The compensatory mutation in *pmrA* makes further emergence of resistance to colistin difficult suggesting that re-introduction of colistin treatment could have been effective for this patient. Finally, the isolate from one patient carried a mutation which conferred modest colistin resistance, without any fitness cost in the absence of the antibiotic. This genotype remains stable in the absence of the antibiotic, and could transmit among patients without being out-competed by strains which are fully susceptible to colistin.

Summary

- ✓ Next-generation sequencing technologies differ amongst each other in terms of sample preparation, the need for amplification and sequencing procedures, the number and length of sequence reads.
- ✓ Data from next-generation sequencers require different algorithms for sequence assembly and alignment, typically based on de Bruijn graphs and Burrows–Wheeler transforms respectively.
- ✓ Draft genomes from next-generation sequencers are comparable in coverage with those from first-generation Sanger sequencers.
- ✓ Next-generation sequencing allows cataloguing of sequence variations across a large number of DNA samples, allowing fine-resolution phylogenetic analysis based on single-nucleotide variations, which help track the origin and transmission of epidemics. It also permits incisive analysis of laboratory evolution experiments.
- ✓ Next-generation sequencing also permits sequencing of community metagenomes and genomes from single bacterial cells.

5 Genome-Scale Analysis of Gene Expression and its Regulation in Bacteria

5.1 Introduction

In the previous chapters, we had discussed methods for studying the genetic content of bacteria on a genomic scale. In the process, we highlighted characteristics of certain bacterial genomes, besides discussing properties common to many bacterial genomes. We also documented intra-specific variation in gene content between bacteria. Different strains of the same bacterial species might show considerable variation in gene content, as a result of mutations getting fixed in a particular context, or the dynamics of gene acquisition and loss. Many of these variations might arise from selection resulting from their niches and/or lifestyles. However, this does not reflect the fact that the same bacterium might be phenotypically distinct under different environmental or cellular contexts. Much of these differences might be attributed to gene expression changes, i.e., under a given condition, a bacterium expresses only a subset of its genes, with the remaining genes being silent. Or, the expression level of the same gene might be quantitatively different between two conditions, in contrast to the dramatic on–off distinction made by the previous statement. Thus, the genetic content of a bacterium can be interpreted by a 'gene expression machinery' in different ways at different times.

The expression of genes, and the manner in which it is regulated, have been studied in recent years using genome-scale techniques. Many of these approaches use DNA microarrays or next-generation sequencing, the basics of which have been covered in the previous two chapters. In the present, core chapter of this book, we will discuss the application of these technologies to the study of gene expression and its control in bacteria. The discussion of techniques and data analysis will focus on gene expression measurements and investigation of genomic regions that bind to a protein of interest. We will also present particular exemplary research as case studies in the process. Novel applications of sequencing technologies towards understanding chromosome conformation will be discussed as part of the case studies. Much of the discussion will be centered around the model organism *Escherichia coli*, with other systems used to highlight certain recent, large-scale genomic studies.

5.2 The process of transcription and the regulation of its initiation: An overview

Transcription is the process by which a DNA sequence is read and used as a template to assemble an RNA sequence.[1] The RNA produced could be the messenger RNA (mRNA), which is a substrate to the ribosome during protein synthesis, and in fact, codes for the amino acid sequence of the corresponding protein. Also produced are non-coding ribosomal and transfer RNA (rRNA and tRNA respectively), which play important roles in protein synthesis. In fact ~90% of all RNA produced by an exponentially growing *E. coli* cell comprises rRNA. Other non-coding RNA with diverse regulatory roles are also produced.

Transcription is essentially an enzymatic process (Fig. 5.1). It takes a DNA sequence, encoding a gene, and free nucleotides as substrates, and produces an RNA sequence as a product. Biologically meaningful transcription starts at specific DNA sites called promoters. It reads the DNA sequence and produces the RNA sequence base-by-base, till it terminates at certain termination sites. The enzyme that catalyses transcription is called the *RNA polymerase* (RPO). The bacterial RPO, broadly speaking, is composed of two major components. The first is called the RPO core-enzyme. The *RPO core-enzyme* itself is a multi-component enzyme, with the following subunit composition–with each subunit encoded by a different gene: α_2, β, β' and ω. The catalytic site of the polymerase is composed of the large β and the β' subunits. The α subunit has two domains: The N-terminal dimerisation domain (α-NTD) helps in the assembly of the two catalytic subunits, and is connected to the DNA-binding C-terminal domain (α-CTD) by a flexible linker. The enigmatic ω subunit appears to be a chaperone for the folding of the β' subunit, and may also interact with regulatory small molecules. The RPO core-enzyme is fully capable of transcription, but does not have the specificity to initiate the process at the correct promoters. This specificity is conferred by the second component of the RPO, called the σ subunit. The complex of the RPO core-enzyme and the σ subunit (or σ-factor) is called the *RPO holo-enzyme*. It initiates transcription specifically from promoter sites. Transcription involves the following broad steps (Fig. 5.1): (a) the formation of the RPO holo-enzyme

[1] Much of the information in this section is drawn from the following excellent review: Browning and Busby. 2004. 'The regulation of bacterial transcription initiation.' *Nature Reviews Microbiology* 2: 1–9.

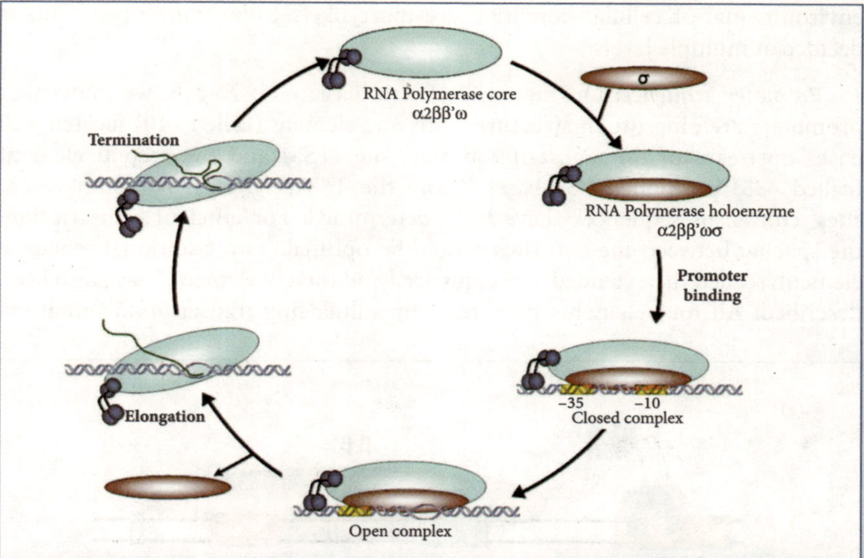

Figure 5.1 *Transcription in E. coli.* This figure shows a simplified representation of transcription in *E. coli*. The multi-subunit core-RNA polymerase is shown by the blue ellipse, with the two copies of the C-terminal end of the α-subunit with its flexible linker represented by the blue dumbbell. The σ-factor is represented by the brown ellipse. The extending RNA chain is shown by the green curve (Elongation and Termination). The two promoter elements are marked by the yellow/orange boxes (Closed complex and Open complex stages). Figure credit: Avantika Lal, National Centre for Biological Sciences, India.

complex by the interaction between the RPO core-enzyme and the σ subunit; (b) the binding of the RPO holo-enzyme to the DNA at promoter sites, forming what is called the *closed complex*; (c) unwinding of the DNA duplex, promoted by the σ subunit, forming the *open complex*; (d) synthesis of the RNA chain base-by-base, its sequence determined by the sequence of bases in the DNA; (e) termination of transcription.

In any bacterial cell, there is considerable difference among genes in their expression levels. Moreover, the same gene can be expressed at different levels under different conditions. In considering the parameters that lead to such variation in transcription and gene expression, it must be noted that the RPO is in short supply in the bacterial cell. A large majority of the RPO holoenzyme molecules are involved in transcribing rRNA genes, thus severely limiting their availability for transcribing protein-coding mRNA and other non-coding RNA genes. Therefore, the distribution of RPO holo-enzyme molecules across promoters is defined such that genes that have important functions in the present environmental or cellular condition are more likely to be transcribed. This is decided at multiple levels:

1 *Promoter sequence*: On the basis of data largely on *E. coli*, we know that promoters are bipartite in structure, with one element (called –10) located ~10 bases upstream of the transcription start site (TSS) and the second element (called –35) positioned ~35 bases before the TSS (Fig. 5.2). For both these sites, consensus sequences[2] have been determined. For efficient transcription, the spacing between the two sites should be optimal. Two additional sequence elements, called the extended –10 sequence,[3] and the UP element,[4] have also been described. All four elements have roles in influencing transcription initiation,

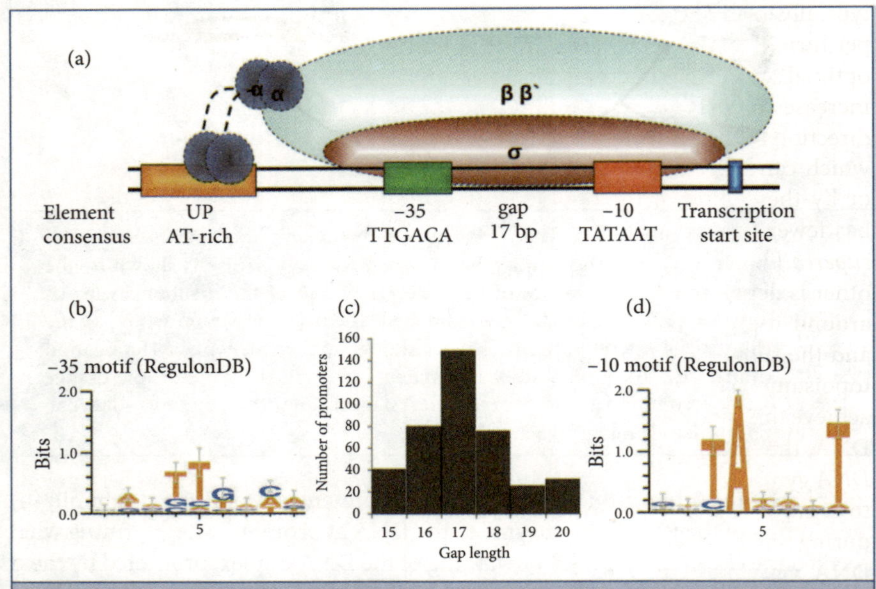

Figure 5.2 *The promoter structure of E. coli.* The major elements of the *E. coli* promoter are shown in (a). The sequence motif for the –35 element is shown in (b), that for the –10 element in (d), and the distribution of the length of the spacer between the two elements in (c). The sequence motifs were drawn using weblogo3 (http://weblogo.threeplusone.com/create.cgi), based on the position weight matrices available in the RegulonDB database (http://regulondb.ccg.unam.mx/). The distribution of spacer lengths was sourced from Schultzaberger et al. 2007. Figure credit: Avantika Lal, National Centre for Biological Sciences, India.

---

2 TATAAT for –10 and TTGACA for –35.

3 The extended –10 element is a 3–4 base sequence immediately upstream of the –10 element.

4 The UP element is a ~20 base sequence, which is intrinsically bent in structure and located upstream of the –35 element.

although the relative importance of each varies from one gene to another. Considering only the −10 and the −35 elements, there is no real *E. coli* promoter that exactly matches the above-defined consensus sequences. The closer a sequence is to the consensus, the stronger it is in attracting the RPO holo-enzyme. Thus, the promoters of *E. coli* show a wide distribution of promoter strengths. Though this sequence difference between promoters will lead to differences in rates of transcription initiation between genes, it is a static property and cannot be modulated by environmental conditions.[5] Therefore, additional players are required to modulate transcription in a condition-dependent manner. The first of these is the topology or the local geometry of the DNA.

2 *DNA supercoiling*: The constraints on the structure of the DNA are such that it typically forms a right-handed double helix, with an optimal number of base pairs per turn. If a sufficiently long linear molecule of DNA is closed into a circle, this optimal structure is maintained. However, twisting of the molecule before closure increases or decreases the number of turns in the molecule, depending on the direction of the twist (Fig. 5.3). This introduces a torsional strain in the molecule, which can be accommodated by changes in the number of base pairs per turn, or by the coiling of the double helix onto itself. Natural DNA predominantly has fewer turns than the relaxed form, a phenomenon which is called *negative supercoiling*. The degree of untwisting of one strand of DNA with respect to the other is defined by the parameter Twist (*Tw*), and the coiling of the double helix around itself by the parameter Writhe (*Wr*). Tw and Wr are interconvertible and the sum of the two is called linking number ($Lk = Tw + Wr$). Two types of topoisomerase enzymes in the cell maintain the supercoiling state of the cell. They achieve this by cleaving one or both DNA strands and allowing free rotation of the DNA, thus modulating the torsional stress on the molecule, and closing the nick. *DNA gyrase* uses ATP to introduce negative supercoils in relaxed DNA, as well as to remove positive supercoils locally introduced by DNA when it is unwinding during replication or transcription. *Topoisomerase I* relaxes negatively supercoiled DNA via a passive process. The supercoiling introduced by these factors in the absence of other DNA binding proteins which bend or wrap the DNA (for example, the nucleosomes in eukaryotes) is called *unconstrained supercoiling*. The difference in ATP requirement between the action of DNA gyrase and that of topoisomerase I immediately links the energy state of the cell with the level of negative supercoiling–the DNA is less negatively supercoiled in stationary phase cells than in exponential phase cells. We note that DNA in *E. coli*–as a whole–is never positively supercoiled, though there is local positive supercoiling induced by the action of motors such as the DNA and the RNA polymerases. However, cellular DNA is bound by proteins which modify the local geometry of the bound DNA, including the RPO at promoters, and therefore, much of the supercoiling

5 Unless prolonged environmental stress leads to the fixation of an appropriately mutated promoter in a cell population.

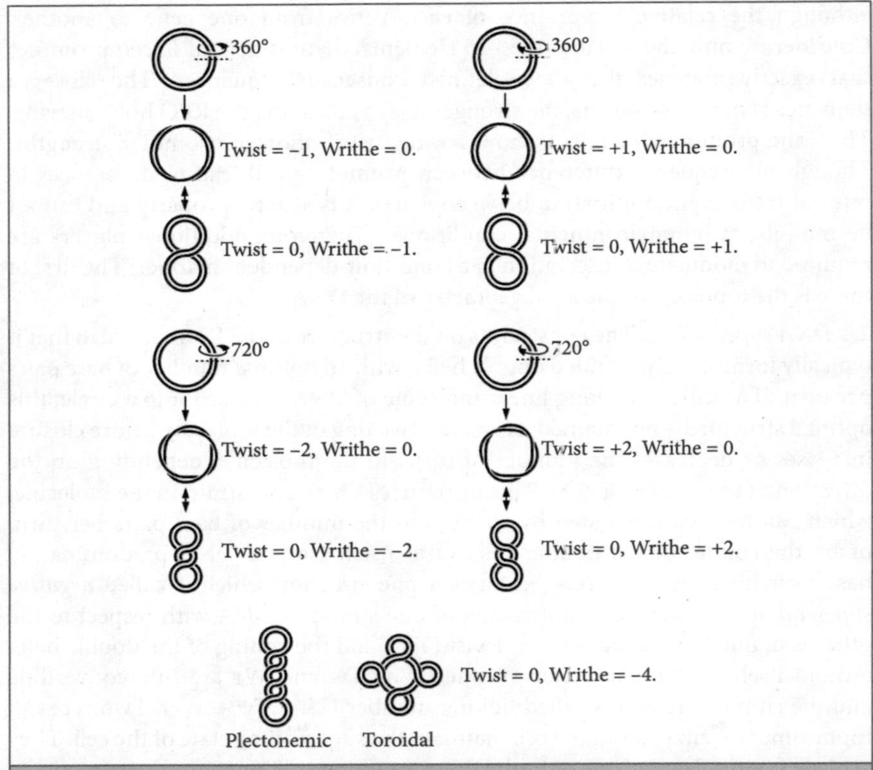

Figure 5.3 *DNA supercoiling*. This figure illustrates four different toy examples for supercoilied DNA. Twist refers to the winding of one strand of the double helix (shown as parallel strands for simplicity here), whereas writhe refers to the winding of the double helix around itself. This figure, authored by Richard Wheeler (Zephyris), is reproduced from Wikimedia Commons under the GNU Free Documentation License (http://commons.wikimedia.org/wiki/File:Circular_DNA_Supercoiling.png).

in the cell is *constrained* by protein binding. The involvement of such proteins in determining the supercoiling state of the cell further makes super-coiling a dynamic property. One can immediately note that changes in the degree of winding of the DNA can influence the binding properties of DNA-binding proteins including the RPO, as well as bring distant parts of the chromosome into close contact, thus affecting processes such as recombination. For further information on supercoiling, readers are referred to other, more focused literature.[6]

---

6 (a) Travers and Muskhelishvili. 2005. 'DNA supercoiling -a global transcriptional regulator for enterobacterial growth?' *Nature Reviews Microbiology* 3: 157–69; (b) Bates and Maxwell. 2005. 'DNA topology.' Oxford University Press.

3 *The competition between multiple σ factors*: As described earlier, the RPO holo-enzyme comprises a multi-subunit core enzyme and a σ factor. Many bacteria encode multiple σ factors (Fig. 5.4). These belong to two broad families. Most known σ factors belong to what is called the σ70 family. In *E. coli*, this family

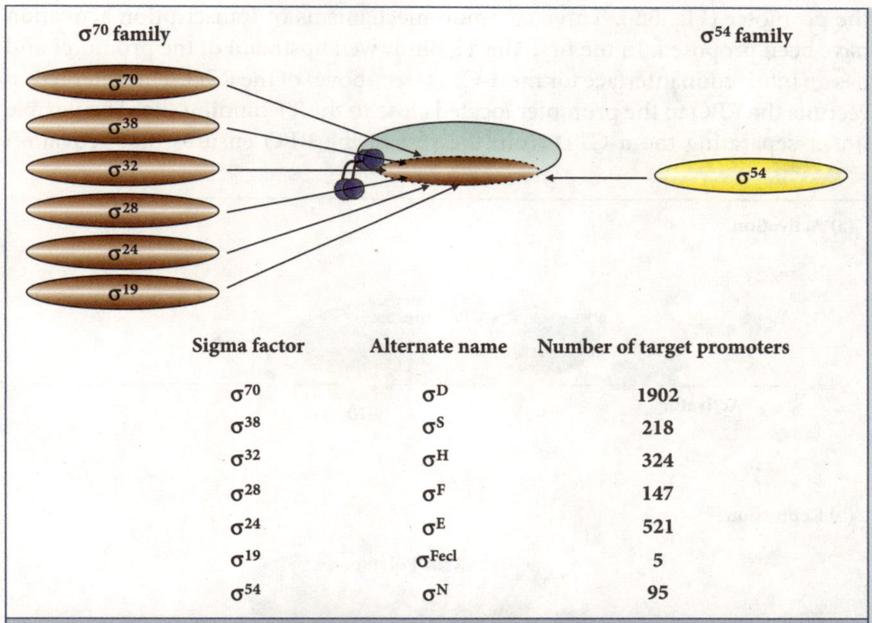

Sigma factor	Alternate name	Number of target promoters
σ^{70}	σ^{D}	1902
σ^{38}	σ^{S}	218
σ^{32}	σ^{H}	324
σ^{28}	σ^{F}	147
σ^{24}	σ^{E}	521
σ^{19}	σ^{FecI}	5
σ^{54}	σ^{N}	95

Figure 5.4 *σ-factors in E. coli.* This figure presents a list of the seven σ-factors encoded by the *E. coli* K12 genome. The number of genes regulated by these σ-factors was derived from the RegulonDB database. Figure credit: Avantika Lal, National Centre for Biological Sciences, India.

includes the major σ factor which is responsible for transcription from a majority of promoters, including those for housekeeping genes. Various other σ factors, regulating specific subsets of genes also come under this family. Studies dealing with this σ factor, and other members of this family, will be discussed in detail later in this chapter. The second σ factor family is called the σ54 family. While transcription initiation by members of the σ70 family involves spontaneous unwinding of the promoter with the σ factor stabilising the unwound form, the promoters under σ54 regulation require unwinding by an active process. *E. coli* encodes one σ54 family member, responsible for transcription under conditions of nitrogen source limitation. σ factors differ from each other in terms of their binding affinities to the core enzyme as well as in the DNA sequences they recognise. There is intense competition among σ factors for binding to the core

RPO.[7] The result of such competition is determined by environmental and cellular conditions, and is a defining factor in the gene expression profile of a bacterial cell.

4 *Transcription factors*: Transcription factors (TFs) are proteins that bind to the DNA and specifically alter transcription of proximal genes. These proteins activate or repress transcription by binding to DNA sites typically located around the promoter (Fig. 5.5). Three common mechanisms of transcription activation have been proposed. In the first, the TF binds well upstream of the promoter and has an interaction interface for the α-CTD (see above) of the RPO. This interaction recruits the RPO to the promoter located close to the TF binding site. The flexible linker separating the α-CTD from the rest of the RPO ensures that activation

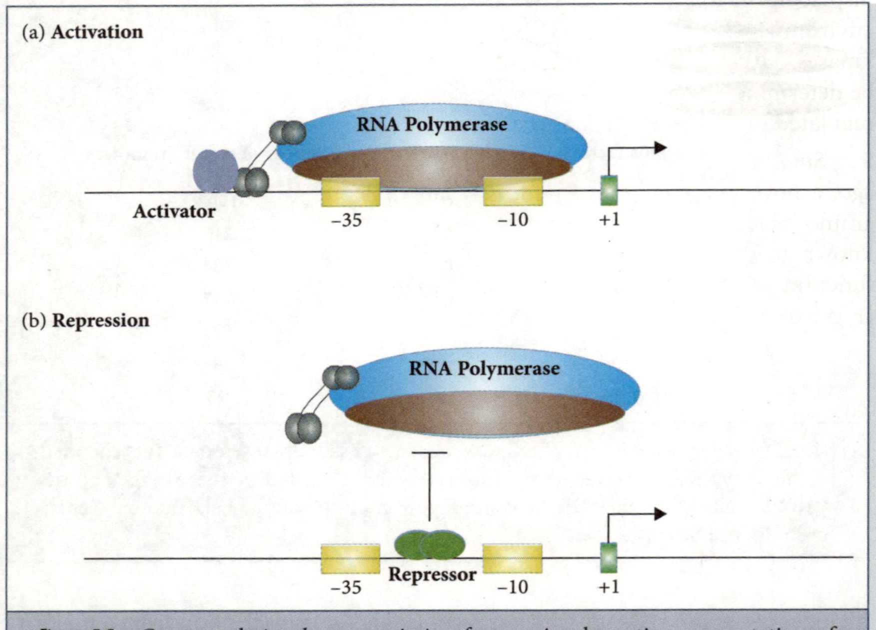

Figure 5.5 *Gene regulation by transcription factors.* A schematic representation of one mode of activation and repression of transcription by transcription factors. (a) The TF (purple ellipse) binds upstream of the promoter and recruits the RPO by interacting with the C-terminal domain of the α-subunit of RPO; (b) The TF (green ellipse) sterically hinders the binding of the RNA polymerase. Figure credit: Avantika Lal, National Centre for Biological Sciences, India

7 (a) Gruber and Gross. 2003. 'Multiple sigma subunits and the partitioning of bacterial transcription space.' *Annual Reviews in Microbiology* 57: 441–66; (b) Österberg S., del Peso-Santos T. and Shingler V. 2011. 'Regulation of alternative sigma factor use.' *Annual Reviews in Microbiology* 65: 37–55.

does not require precise positioning of the TF binding site relative to the promoter. In the second mechanism, the TF binds close to the promoter and recruits the RPO by protein–protein interactions with the σ factor. The third mechanism operates at certain promoters where the linear spacing between the −10 and the −35 promoter elements is sub-optimal. These TFs rotate the DNA so that the two sites are positioned in a manner permissive to RPO binding. Similarly, three general mechanisms for repression have been formulated. The first involves direct steric hindrance of RPO binding by TF–DNA interactions. The second mechanism involves TF binding to distal sites on the DNA and looping the DNA separating the two sites. The two sites are generally located so that the looping occludes RPO binding, thus preventing transcription. The third mechanism involves a repressor displacing an activator. The activities of TFs are regulated by direct binding of an environmental or a cellular signal molecule, or by phosphorylation by receptor kinases. TFs may also be regulated by their expression levels, which in turn could be determined by higher level TFs or DNA supercoiling, or other players such as regulated proteolysis or sequestration by protein–protein interactions.

5 *Small molecules that interact with the RNA polymerase*: Over four decades ago, a novel nucleotide was found to be abundant in *E. coli* when grown under amino acid starvation. This molecule, the 'alarmone' (p)ppGpp,[8] is now known to directly bind to the RPO and affect transcription. A characteristic function of this molecule is to decrease transcription of ribosomal RNA (Fig. 5.6). We recollect here that rRNA synthesis keeps most of the RPO molecules

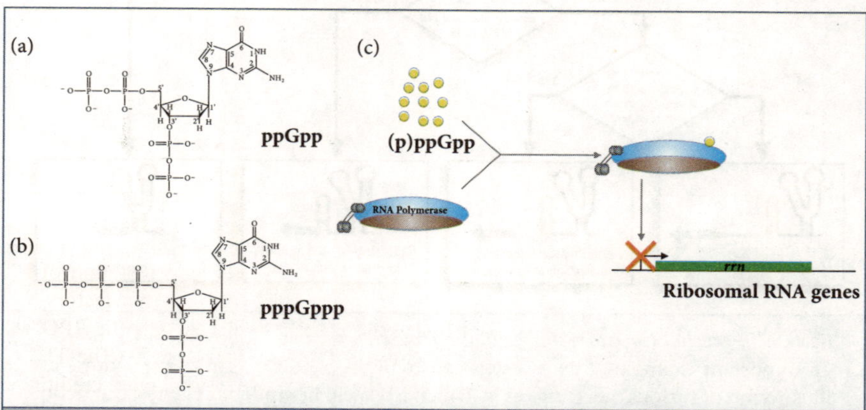

Figure 5.6 *Gene regulation by the small molecule-alarmone ppGpp.* This figure shows the structures of (a) ppGPP and (b) pppGpp. (c) In the canonical mode of action of (p)ppGpp, the small-molecule binds to the RPO – in the ω-subunit – and inhibits the transcription of ribosomal RNA operons.

[8] Ganosine tetra phosphate and Guanosine penta phosphate.

busy, and will therefore be expensive and futile under nutrient limiting conditions when substrates for protein synthesis and other macromolecular biosynthetic processes are not available. Though there is much knowledge on characteristics that determine how a promoter responds to ppGpp, much remains to be worked out in the field.

6 *Regulation beyond transcription initiation–regulatory RNA and elongation/termination factors*: RNA itself can be a regulator of gene expression, if not during transcription initiation, then by controlling transcription termination, mRNA stability and protein synthesis. Regulatory RNA molecules might act in *cis* or in *trans*. Many *cis* regulatory structures are part of the mRNA itself, and form secondary structures around the 5′-UTR[9] of the transcript. These structures might

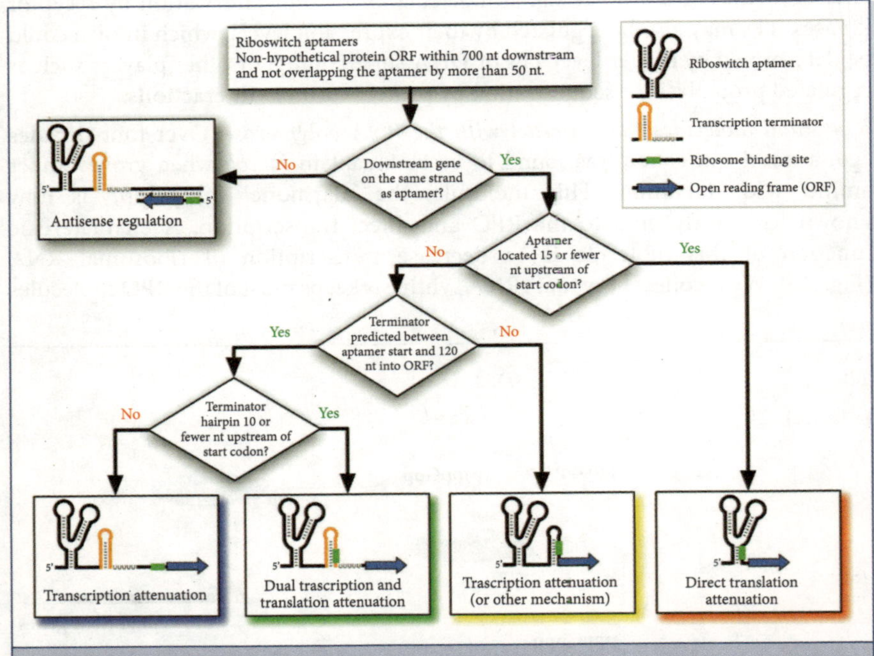

Figure 5.7 *Mechanisms of gene regulation by riboswitches.* Riboswitches represent an important example of gene regulation by RNA modules. This figure shows an illustration of various mechanisms by which riboswitches influence gene expression. This figure is reproduced under the Creative Commons Attribution license from Barrick and Breaker. 2007. 'The distributions, mechanisms, and structures of metabolite-binding riboswitches.' *Genome Biology* 8: R239. © Barrick and Breaker 2007; licensee Biomed Central.

9 5-Untranslated Region refers to the sequence of nucleotides between the transcription initiation site and the ORF start site.

terminate transcription prematurely so that functional mRNA is not produced, or they might block the ribosome from initiating protein synthesis. The formation of such structures is generally dependent on cellular conditions. Some structures form at lower temperatures, and others form depending on the availability of small molecules to which they can bind (these are called riboswitches) (Fig. 5.7). *cis* or *trans* antisense RNA could be produced, which on annealing to their target mRNA can block translation and/or activate pathways for double-stranded RNA degradation. These require a certain level of sequence complementarity between the regulatory RNA and the target mRNA. In the case of *trans*-encoded anti-sense RNA, where the complementarity is low, RNA chaperones such as the protein Hfq are required. Besides such RNA molecules, various transcription elongation and termination factors such as GreA/B, NusA/B and Rho may also be regulatory, affecting gene expression of specific genes.

5.3 Measuring gene expression on a genomic scale: Technologies

Rapid progress in reading the complete sequences of genomes generated much interest in determining the gene expression profile (or the transcriptome) of these genomes, rapidly and with reasonable accuracy. By a gene expression profile, we mean the organisation and abundance of each transcript encoded by the genome under a panel of environmental and cellular (including genetic) conditions. Traditionally, transcripts that are expressed in a cell were described using the labourious process of generating cDNA clones and sequencing small fragments, referred to as expressed sequence tags. However, the nature of this experimental approach did not lend itself to rapid characterisation of gene expression profiles on a truly genomic scale in a time-critical manner. Therefore, alternative methods had to be developed and evaluated.

The first truly genome-scale approach for gene expression analysis, which became popular, was the use of DNA microarrays. Just as hybridisation of genomic DNA to a suitably designed microarray could enable a study of variation in gene content across related bacteria, hybridisation of cDNA–generated from RNA isolated from cells grown under appropriate conditions–to a microarray enabled descriptions of gene expression profiles. Most microarray-based gene expression studies used probes designed to interrogate previously annotated genes. Later, the development of genome-tiling microarrays, which are annotation un-aware and require only the genome sequence of the target organism, paved the way for describing novel transcripts, and permitted what is called experimental genome annotation. However, these approaches have their limitations. As with comparative genome hybridisation, transcriptome analysis using microarrays required information about the genome sequence, and could not be applied to an organism whose genome sequence was not known. For many users, the application of microarrays was limited to interrogating gene expression from genomic regions probed in commercially-available array design–custom design of a microarray remained challenging to the non-specialist as a result of the mind-boggling arrays of physicochemical and statistical considerations that went into the process. Further, fluorescence intensities measured from microarrays saturated at high levels, thus limiting the dynamic range of measurement. The sensitivity of many microarrays to low expression levels was also questionable. Added to these difficulties was the range of statistical methods that were required to process the data (Chapter 3). As a possible consequence of these factors, the acceptability of microarrays remains low among many biologists. That said, microarrays have proved to be great tools in generating invaluable data leading to novel findings, and some of these studies will be discussed in this chapter.

As next-generation sequencing technologies emerged, it became apparent that the high depth of coverage made them at least as quantitative as microarrays for gene expression measurements and other allied applications. The fact that the coverage or depth of sequencing was determined by the discretion of the experimentalist (subject to financial and logistical considerations) meant that the dynamic range in measurement was less of an issue than it was with microarrays. One could use these technologies to assemble the gene expression profiles of organisms without a reference genome sequence, although one might argue about the utility of this to bacterial systems where obtaining a reasonable draft genome sequence using next-generation sequencing is relatively routine. For complex eukaryotes, these technologies also offered the possibility of discovering novel transcript isoforms. Though some aspects of computation essential for interpreting microarray data could be dispensed with while analysing next-generation sequencing data, these data present their own challenges in data storage and processing pipelines in going from raw sequence reads to measures of gene expression. These will be discussed below, as will be several sequencing-based studies of bacterial gene expression and experimental genome annotation.

Both microarrays and next-generation sequencing approaches require invasive techniques in sample processing: Cell lysis followed by RNA purification. This made gene expression analysis at high temporal resolution difficult to achieve. Such analysis require live-cell measurement techniques. This has been achieved using libraries of strains carrying promoters fused to fluorescent reporters. A genome-scale analysis of gene expression at this level requires automation in parallel processing of thousands of strains each carrying a promoter–reporter fusion. Thus, these approaches are not readily available to researchers interested in relatively obscure organisms.

Many of the analysis techniques applied to microarrays have been presented in Chapter 3 and will not be revisited here; novel analysis methods–where relevant–will be presented along with the case studies. The following section will detail computational approaches for transcriptome analysis using next-generation sequencing.

5.4 Next-generation sequencing for gene expression measurements: Data analysis

Sequencing is generally thought to be an operation performed when it is needed to know the string of bases that make up a nucleic acid sequence. Many look askance when presented with the idea that one can use sequencing to 'quantify' one nucleic acid sequence relative to another. In fact, the idea is simple and intuitive. In order to achieve this, all one has to do is to have a system that sequences a nucleic acid *n* times, such that *n* is in some way proportional to the amount of that sequence present in a sample. It is also important to note that if one wants to quantify a single nucleic acid sequence, this is not the way to go, unless one is prepared to perform careful experiments with spike-in controls (it is a different matter altogether that nobody would really want to use a next-generation sequencer to do this, unless a specific application requires very high coverage of a sequence for whatever reason). For a typical RNA-seq experiment, one needs a frame of reference, i.e., one can state to a reasonable degree of approximation that the amount of one nucleic acid sequence is *x* times that of another. This is because, the total number of sequence reads that one can get out of a sequencer is dependent on the technology and on the amount of money that can be spent by the investigator on multiple sequencing runs. However, among this total number of reads, the fraction coming from one sequence will be proportional to the relative abundance of that nucleic acid sequence in the sample. We will henceforth use the term 'semi-quantitative' to express the numbers that one can obtain from next-generation sequencing when quantifying nucleic acids. It is also worth noting here that similar constraints apply to measurements using DNA microarrays.

Let us now establish the fact that next-generation sequencing can be semi-quantitiative. *E. coli*, when grown in rich Luria broth, has a population doubling time of 20–30 minutes in the exponential growth phase. However, DNA replication, limited by the speed of the DNA polymerase, needs an hour or so to complete. This is a conundrum–on average, the population doubles in a time that is much shorter than what it takes to replicate the chromosome. The solution to this problem comes from the fact that the origin of replication fires multiple times in an average *E. coli* cell, so that one replication fork follows another.

Thus, in an average culture doubling exponentially, one can expect DNA around the origin of replication to be more abundant than that around the terminus. However, this will not be the case in a culture during the stationary phase. One could isolate genomic DNA from a mid-exponential and a stationary phase *E. coli* population and subject it to next-generation sequencing to test if the above phenomenon is reflected in the sequencing data. We align the sequencing reads from our experiment to the reference *E. coli* genome so that we now know which part of the genome each read originated from.[10] This alignment also allows us to calculate–for each base position on the genome–the number of mapped reads, which[11] should be proportional to the abundance of that sequence in the sample. We can plot the number of reads mapped (coverage) on the *y*-axis against the chromosomal coordinate on the *x*-axis. Such a plot for the exponential phase culture shows that the coverage is higher around the origin of replication, but declines smoothly towards the terminus (Fig. 5.8). In contrast, the coverage plot is relatively flat for the stationary phase sample. One of the first applications of next-generation sequencing to gene expression measurements showed the presence of a linear correspondence between the concentration of spiked-in RNA of known quantities and the number of reads that mapped to them (Fig. 5.8).[12] These examples should be a reasonably convincing evidence that deep sequencing can be semi-quantitative.

The next question is how do we apply data from next-generation sequencing to the quantification of gene expression? The basic requirement for such experiments is RNA isolation from the cells of interest, followed by cDNA synthesis (except in recent direct RNA sequencing applications), followed by library preparation and sequencing. Next-generation sequencing-based techniques for gene expression analysis are generally referred to as *RNA-seq* experiments. Similar to data from microarrays, data from sequencing require a certain level of data processing. We will consider these in some detail below. However, for now, let us take a step back and ask the question: Can one use short read next-generation sequencing data to first *identify* transcripts which are expressed, describe their characterisitics and thereby achieve experimental genome annotation? At this point, we are not necessarily interested in quantifying the abundance of transcripts, but only in identification. The answer is yes, and the class of approaches applicable to this question is called *transcriptome assembly*, which we will discuss below. Material for this section on transcriptome assembly primarily refers to an excellent review on the subject by Martin and Wang.[13]

---

10 We neglect reads that map to repetitive loci.

11 Subject to sampling and systemic errors.

12 Mortazavi A., Williams B. A., McCue K., Schaeffer L. and Wold B. 2008. 'Mapping and quantifying mammalian transcriptomes by RNA-seq.' *Nature Methods* 5: 621–28.

13 Martin and Wang. 2011. 'Next-generation transcriptome assembly.' *Nature Reviews Genetics* 12: 671–82.

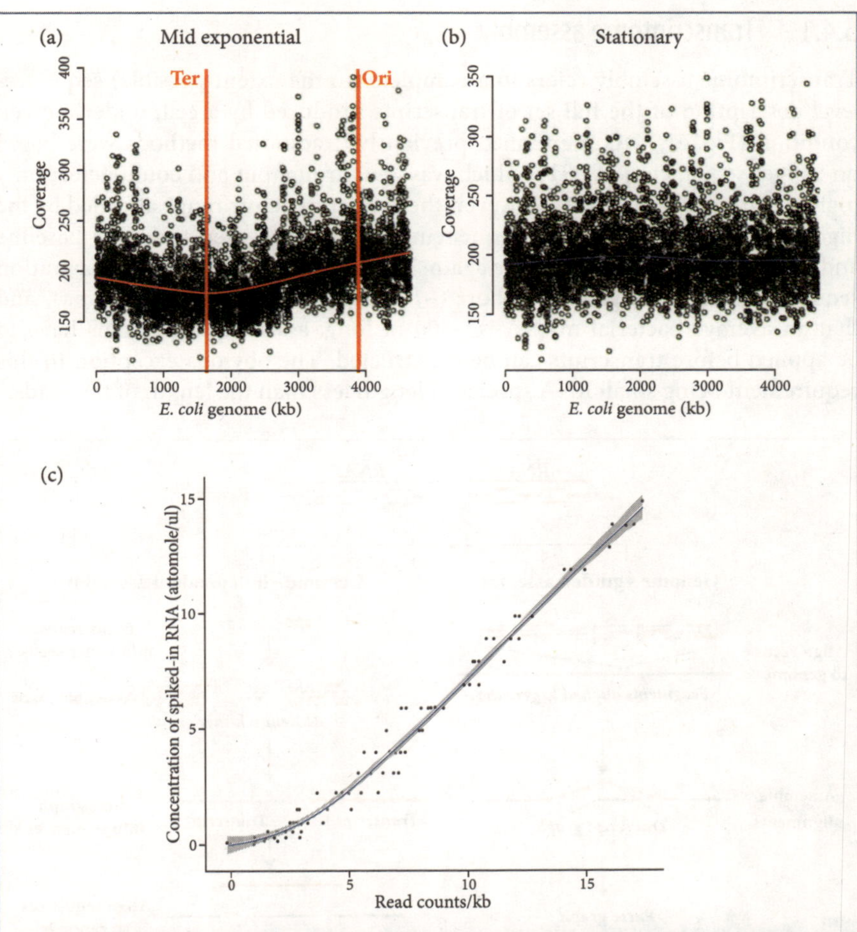

Figure 5.8 *Quantitative nature of deep-sequencing data.* This figure shows the sequencing coverage (*y*-axis) as a function of base position (*x*-axis) along the *E. coli* chromosome during (a) exponential phase; (b) stationary phase. The exponential phase sample was taken when the estimated average population doubling time was slightly higher than the time required by an *E. coli* cell to replicate the chromosome and divide, resulting in multi-fork replication in a sub-population. This is reflected in the plot for the exponential phase, where the coverage is slightly higher around the origin of replication (Ori) than around the terminus (Ter), whereas it is relatively flat during stationary phase. These are from data generated in our lab by Aalap Mogre. (c) This panel shows the correlation between read counts (per-kb) and the concentration of spiked-in RNA molecules on a log-log scale. These are from data provided by Hardik Gala (Institute of Stem Cell Biology and Regeneration, Bangalore).

5.4.1 Transcriptome assembly

Transcriptome assembly refers to a complete (to the extent possible) sequence-level description of the full set of transcripts produced by a cell, under a given condition (Fig. 5.9). As mentioned previously, traditional methods were based on Sanger sequencing of ESTs, which was low throughput and could detect only highly expressed transcripts. However, the greater dynamic range afforded by the high depth of next-generation sequencing allows us to even trawl the baseline and identify rare transcripts. Once again, because of the fact that next-generation sequencing reads are generally short (~100 nt good quality for Illumina), and that an average bacterial mRNA is ~800 nt long, assembly algorithms have to be applied before transcripts can be constructed. The obvious exception to this requirement being small RNA species of length less than the length of the reads.

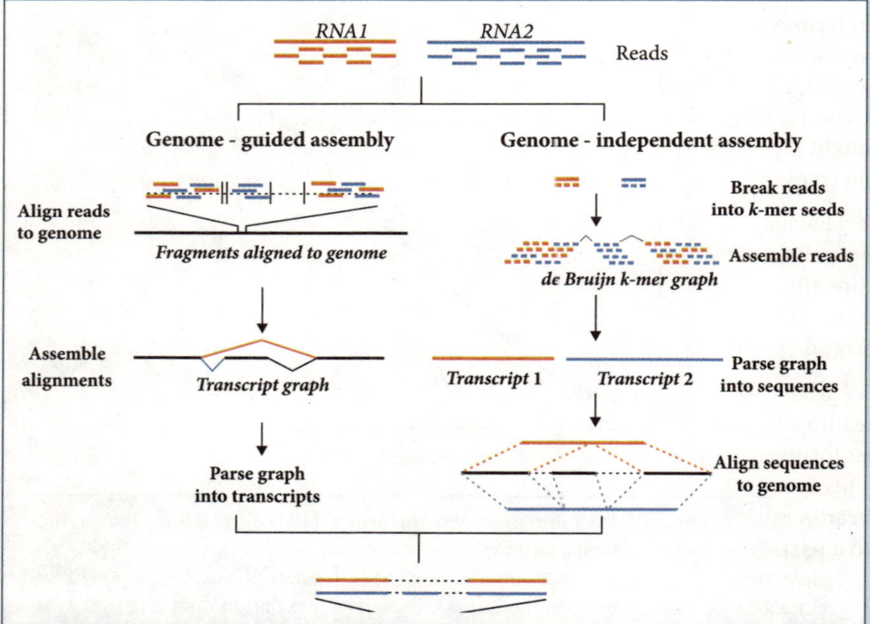

Figure 5.9 *Transcriptome assembly.* This figure shows a schematic representation of a transcriptome assembly pipeline, by which short reads from an RNA-seq experiments are assembled together into transcripts, either aided by a reference genome (left; genome-guided assembly), or *de-novo* (right; genome-independent assembly). Alignment of *de-novo* assembled sequences to a genome is optional and typically used in hybrid assemblies described below. Figure credit: Avantika Lal, National Centre for Biological Sciences.

5.4.1.1 Why is transcriptome assembly not the same problem as genome assembly?

Many of the computational challenges of assembling genomes from short reads apply to transcriptome assembly. The fundamental challenge is to assemble millions of short reads, amounting to gigabases of data, into a much smaller dataset of contigs (in the genome assembly parlance) or transcripts (in the present case). While many programs such as Velvet and Allpaths have been successful in assembling genomes, they cannot be directly applied to transcriptome assembly. This is because of properties unique to transcriptome sequencing, when compared to genome sequencing:

1 The depth of coverage in a genome sequencing dataset is relatively uniform across the length of the genome. However, transcript levels differ across genes by several orders of magnitude; therefore genes differ dramatically from each other in terms of their depth of coverage. For many genome assemblers, coverage is an important factor in determining repetitive regions. This means that these programs would call highly expressed genes as repetitive and therefore be problematic. Genome assemblers also tend to discard loci of low depth of coverage, as these might represent sequencing errors. This means that such programs when acting on transcriptome data would discard transcripts with low expression levels.

2 Each transcript arises from one strand of the genomic DNA. There could be pairs of transcripts that overlap each other–for example, in the anti-sense direction.[14] Then it could be difficult to disentagle the two as separate transcripts. However, this can be managed using experimental techniques which allow for strand-specific sequencing.

3 Many organisms–especially higher eukaryotes–undergo mRNA splicing leading to multiple transcript isoforms emerging from the same gene. Many isoforms share common exons, an aspect that does not concern genome assembly. This is not a major concern in bacterial research, but is nevertheless an issue to bear in mind while assessing the utility of an analysis pipeline (software packages) to a particular transcriptome dataset.

5.4.1.2 Considerations in the design of a transcriptome (assembly) experiment

There are various issues that need to be considered before conducting a transcriptome assembly project:

1 Much of the interest in transcriptomes is focused on mRNAs (and small non-coding RNAs), which form a small minority of the total RNA population of a cell. Therefore, one might have to enrich this subpopulation of RNAs. In eukaryotes, this is facilitated by the fact that most mRNAs are naturally tagged at their 3′-ends

[14] *Transcript A* from the + strand, and *Transcript B* from the – strand being expressed from different promoters.

by a poly-A sequence. This luxury does not exist for bacterial mRNA.[15] Thus, in bacteria, methods for depletion of the majority rRNA molecules are used. One has to consider the possibility that such interventions might bias even the mRNA population towards certain sequence types. A final judicious call must be made taking these factors into account, along with the depth of coverage required for the mRNA population, and the availability of resources for achieving the required coverage without compromising data quality and reliability.

2 A second consideration in sample preparation is the use of PCR amplification during library construction, which could lead to preferential amplification of certain sequence stretches and therefore uneven coverage even among segments of the same gene. Single molecule sequencing approaches, such as the technology developed by Helicos, do not require amplification and therefore get around this problem by default.

3 Finally, the question of which sequencing technology/approach to use arises. For most bacterial applications, short read technologies are sufficient. Lower coverage, long-read technologies such as 454 pyrosequencing allow for better assemblies, but are not as effective in quantifying gene expression as the short-read, high coverage technologies. The limitations of short reads can be overcome by paired-end sequencing, where the constraint of insert length permits better isoform resolution. Again the need for these approaches is likely to be minimal in bacterial studies.

5.4.1.3 Reference-assisted transcriptome assembly

Transcriptomes can be assembled *de-novo*, with only the short-read sequences as the input, or with the help of a reference genome sequence. In this section, we will discuss the latter.

What is the need for reference-assisted assembly (Fig. 5.9). In organisms where RNA splicing occurs, leading to multiple isoforms of the same gene, reference-assisted assembly helps in ascertaining which distant portions (within reasonable limits) on the genomic sequence come together to form a transcript. Reference-assisted assembly, as defined here, might be superfluous for work in bacteria, where splicing is uncommon.[16] However, such assemblies can help define transcript boundaries in bacteria–although one could achieve this using other methods (see case studies on experimental genome annotation). That said, assemblies enable discovery projects, and might lead to the identification of unanticipated splicing events of functional importance even in bacteria. Therefore, there is value in understanding these techniques.

---

15 Though the reader is encouraged to refer to Sidney Kushner's work on poly-A tails in bacterial RNA.

16 Exceptions being self-splicing auto-catalytic group I and group II introns in a few genes in bacteria, bacteriophages, mitochondria and chloroplasts.

The first requirement for reference-assisted assembly is a set of good alignments of sequencing reads to the reference genome. Where RNA splicing occurs, reads spanning the junctions between two exons will align poorly to the reference genome. Therefore, to make best use of reference-assisted assembly, the alignment tool should be able to split unmapped reads and align individual segments to the genome. If two segments of the same read align to two different portions of the reference genome, then it is likely that the intervening genomic locus is an intron that is spliced out in the mature transcript. Alignment pipelines with the above ability are referred to as being *splice-aware*; examples include tools such as TopHat, SpliceMap and GSNAP. Once these alignments are generated, assembly is performed. Since each transcript is expected to cover only a small portion of the genome, assemblies are constructed independently across many segments of the genome. This reduces the size of each assembly by several orders of magnitude and multiple assembly operations can be carried out in parallel. Similar to genome assemblies, overlapping reads from each assembly segment are lined up. These overlaps are converted into graphs–one for each assembly segment–through which paths can be traced to identify all possible isoforms emerging from that assembly segment. Two commonly used programs are *Scripture* and *Cufflinks*, which differ from each other in the manner in which the graphs are constructed and then traversed. In *Scripture*,[17] every base is a node, and two bases are connected if a read joins them. Then all possible paths through this graph, for which the coverage is over a threshold, are determined and listed as possible transcripts. *Cufflinks*[18] connects genomic *fragments*[19] together if they are 'compatible'. Two distal regions on the genome become linked if the same read connects the two together; two reads are connected together if they overlap. For paired end sequencing, a pair of fragments, emerging from two sets of read pairs are deemed compatible if they can be explained by the same transcript. Using such rules, an assembly is built such that every fragment is associated with a transcript, and is tiled by reads. The algorithm also ensures that the assembly represents the minimum set of transcripts so that every fragment is associated with an assembled transcript. Thus, *Cufflinks* takes a conservative approach towards assembly, whereas *Scripture* produces a more exhaustive list of possibilities.

If a good quality reference genome is available, reference-assisted assembly offers a sensitive option for identifying transcripts and their boundaries. Since

---

17 Guttman M., Garber M., Levin J. Z., Donaghey J., Robinson J., Adiconis X., Fan L., Koziol M. J., Gnirke A., Nusbaum C., Rinn J. L., Lander E. S. and Regev A. 2010. 'Ab initio reconstruction of cell type–specific transcriptomes in mouse reveals the conserved multi-exonic structure of lincRNAs.' *Nature Biotechnology* 28: 503–10.

18 Trapnell C., Williams B. A., Pertea G., Mortazavi A., Kwan G., van Baren M. J., Salzberg S. L., Wold B. J. and Pachter L. 2010. 'Transcript assembly and quantification by RNA-Seq reveals unannotated transcripts and isoform switching during cell differentiation.' *Nature Biotechnology* 28: 511–15.

19 *Fragment* refers to the cDNA piece whose ends are described by sequencing reads.

the genome is split into several loci and independent assemblies performed for each, the computational memory requirements are minimal, when compared to a *de-novo* assembly. It also has the disadvantage in complex genomes, where transcripts skipping large introns will be missed.

5.4.1.4 *De-novo* transcriptome assembly

The idea behind a *de-novo* transcriptome assembly is the same as that behind genome assembly: To stitch together reads into transcripts without the aid of a reference genome. As mentioned earlier, differences between a genome and a transcriptome assembly arise because of the large variation in depth of coverage, and the sharing of the same exon among many transcripts–characteristics unique to transcriptomes. Many de Bruijn graph-based genome assemblers (see Chapter 4) have been enhanced with special modules applicable to transcriptome data. These include ABySS and Velvet, whose transcriptome assembly variants are called transABySS and Oases. Other software such as Trinity have been developed primarily for transcriptome assembly. Though these programs have introduced several novelties to the assembly pipeline, one particular concept that has emerged is the need for the use of multiple *k*-mers in assembling transcriptomes. Readers might remember that *k* is a parameter in a de Bruijn graph-based assembly process, which determines the degree of overlap between reads that is required for them to be linked together, and therefore, the sensitivity and specificity of the assembly. Higher the value of *k*, less likely is an overlap between two reads. Therefore, longer *k*-mers are better suited to assembling transcripts with high depth of coverage. Shorter *k*-mers are required to assemble transcripts which are expressed at low levels, and therefore, show less coverage. *De-novo* transcriptome assembly is hence generally performed at multiple *k*-mers. These assemblies are then merged together to produce a final transcriptome assembly. Both transABySS[20] and Oases[21] adopt this technique, whereas Trinity[22] makes use of an entirely new approach–also based on de Bruijn graphs–for assembly.

---

[20] Robertson G., Schein J., Chiu R., Corbett R., Field M., Jackman S. D., Mungall K., Lee S., Okada H. M., Qian J. Q., Griffith M., Raymond A., Thiessen N., Cezard T., Butterfield Y. S., Newsome R., Chan S. K., She R., Varhol R., Kamoh B., Prabhu A. L., Tam A., Zhao Y., Moore R. A., Hirst M., Marra M. A., Jones S. J., Hoodless P. A. and Birol I. 2010. 'De novo assembly and analysis of RNA-seq data.' *Nature Methods* 7: 909–12.

[21] Schulz M. H., Zerbino D. R., Vingron M. and Birney E. 2012. 'Oases: Robust de novo RNA-seq assembly across the dynamic range of expression levels.' *Bioinformatics* 28: 1086–1092.

[22] Grabherr M. G., Haas B. J., Yassour M., Levin J. Z., Thompson D. A., Amit I., Adiconis X., Fan L., Raychowdhury R., Zeng Q., Chen Z., Mauceli E., Hacohen N., Gnirke A., Rhind N., di Palma F., Birren B. W., Nusbaum C., Lindblad-Toh K., Friedman N. and Regev A. 2011. 'Trinity: Reconstructing a full-length transcriptome without a genome from RNA-Seq data.' *Nature Biotechnology* 29: 644–52.

De-novo transcriptome assembly is useful when a reference genome is not available. Even in situations where a reference genome is available, but is of a low quality, *de-novo* assembly can be useful in recovering loci not covered by the genome assembly. In higher eukaryotes, *de-novo* assembly gives the opportunity to identify splicing between distal parts of the chromosome (referred to as trans-splicing); it may be noted again that reference-assisted assembly is performed over shorter segments of the genome thus eliminating the discovery of trans-splicing events. On the flipside, *de-novo* assembly requires much higher computer memory and a much higher depth of coverage than reference-assisted assembly to achieve good results.

In summary, in the context of bacterial studies, *de-novo* assembly can be of use where the genome assembly is incomplete.

5.4.1.5 Combining reference-assisted and *de-novo* assemblies

Both assembly methods have their own particular strengths. Reference-assisted assembly algorithms are highly sensitive, whereas *de-novo* assemblies can detect novel, unexpected transcript isoforms. Combining the two methods could offer the benefit of leveraging their respective strengths.

In cases where a reference genome is available, reads could first be aligned to it, following which unmapped reads could be assembled *de-novo*. Many *de-novo* assemblers support combinations of short and long reads. Here, assemblies derived from the reference-assisted stage can be treated as long reads and fed as input–alongside the unmapped reads–for the second, *de-novo* assembly stage. In situations where the reference genome is of dubious quality, *de-novo* assembly can be performed first. The longer contigs thus obtained can be aligned more accurately to the reference genome. These alignments can be combined, thus permitting scaffolding of the fragmented transcripts that *de-novo* assembly typically produce. Contig/scaffold alignments can also be combined with mapping of unassembled reads to the genome to help extend them further.

5.4.1.6 Assessing assembly quality

How does one assess the quality of an assembly? This is an open problem. Martin and Wang[23] have listed five intuitive measures of quality. It is to be noted here that these require a gold standard set of reference transcripts against which assemblies can be computed. These measures are:

1 ***Accuracy***: The percentage of bases in the assembly which match the corresponding reference transcript sequence.

---

[23] Martin and Wang. 2011. 'Next-generation transcriptome assembly.' *Nature Reviews Genetics* 12: 671–82.

2 ***Completeness***: The percentage of the total reference transcript length that is covered by the assembly.

3 ***Contiguity***: The percentage of the reference transcript length covered by a single-longest assembled transcript.

4 ***Variant resolution***: The percentage of the known isoforms recovered by the assembly for a reference gene.

5 ***Occurrence of chimeras***: The possibility that certain assembled transcripts are fusions of distinct reference genes. We note here that some of these chimeras could be real transcripts. However, real and erroneous chimeras can be distinguished by comparing the depth of coverage of reads spanning the chimeric junction and comparing it with the coverage over other segments of the transcript.

It is anticipated that with further development of next-generation long read technologies such as SMRT sequencing from Pacific Biosciences, transcriptome assembly will become redundant.

5.4.2 Measuring gene expression levels

5.4.2.1 From read counts to measures of gene expression

Once a list of genes or transcripts (in this section, we will use *genes* to refer to both genes and transcripts) is available–either through careful annotation of genomes or by experiment-guided techniques such as transcriptome assembly–one can derive approximations of the expression levels of genes and compare these measures across different environmental and/or genetic conditions. This again requires reliable mapping of reads to the genome or to the transcriptome of interest. The next step is to obtain a number for each gene, which is proportional to its expression level. This is typically the number of reads which map to the gene of interest. However, calculating this measure of expression such that it is readily suitable for comparisons across samples is not as straightforward as it sounds. Why? Although the various technical issues necessitating extensive data processing for microarray experiments do not bother transcriptome sequencing data, other complications emerge. First, in studies comparing the gene expression levels of genes from multiple samples, the sequencing *library size*–i.e., the total number of reads obtained by sequencing that sample–varies across samples. Therefore, a simple count of the number of reads that map to a gene cannot be directly used. Aditionally, there could be variability in expression levels within a sample because of differences between genes in their lengths and G+C content. Since fragmentation of cDNA is random, one could expect longer genes to be fragmented more times than shorter ones. Preferential sequencing of regions with particular G+C ranges could arise from biases inherent to PCR amplification. Finally, sets of very highly expressed genes could be unique to one condition but not the other. This affects the overall distribution of the sequencing reads across all the genes in the sample. These effects mean that certain normalisation procedures should be applied to RNA-seq data as well (Fig. 5.10). To all the methods that we describe below, the input is a matrix where each column represents a sample, each row, a gene, and each cell, the raw number of reads mapping to that gene in a given sample. We call this number the raw read count (*RC*). We also note here that understanding and dealing with sources of artifacts in RNA-seq experiments is a work in progress, and there can be many exceptions to anything that may be construed as a rule from the discussion below.

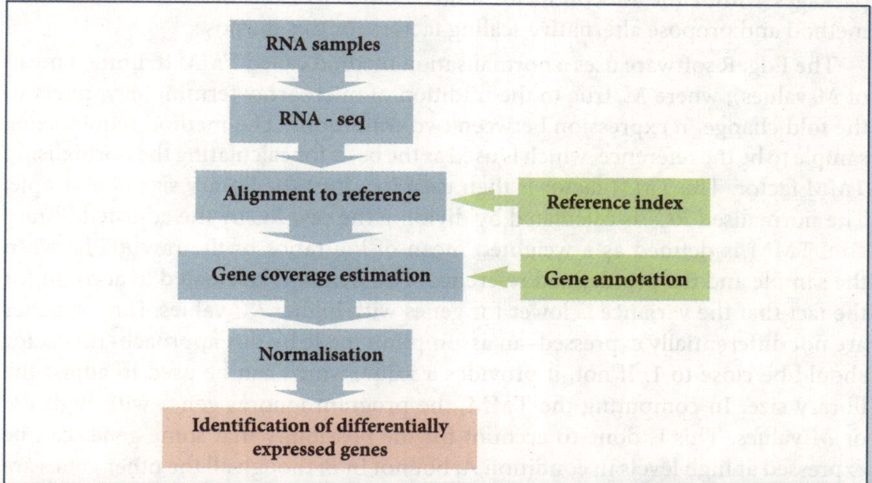

Figure 5.10 *Gene expression quantification from RNA-seq experiment.* This figure shows a pipeline for analysing RNA-seq data towards identifying genes that are differentially expressed between two conditions. Multiple methods are available for normalisation and for defining differentially expressed genes based on sound statistics; these are discussed in the main text.

The most popular method of normalisation, first used in some of the earliest RNA-seq studies, is the measure called *RPKM* (reads per kilobase per million).[24] For each gene, the number of mapped reads is first divided by the length of the gene and then the library size (i.e., the total number of reads obtained for that sample). This accounts for the expectation that longer genes are more likely to have a mapped read than shorter ones. It also accounts for differences between samples in the library size. Most transcriptome studies are interested in comparing the expression levels of the same gene across multiple conditions. In such cases, division by gene length is not required. This is an important point as accounting for gene length by division may, strangely enough, not be appropriate: A recent

[24] Mortazavi A., Williams B. A., McCue K., Schaeffer L. and Wold B. 2008. 'Mapping and quantifying mammalian transcriptomes by RNA-seq.' *Nature Methods* 5: 621–28.

survey[25] of normalisation procedures for RNA-seq data, observed that different datasets show different relationships between *RC* and gene length; and in cases where there is a positive correlation between *RC* and gene length, division of the former by the latter appears to be insufficient to remove the bias. Therefore, newer methods (see below), which are directed at preparing RNA-seq data for computing differential expression of the same gene across multiple conditions, do away with normalisation for gene length.

Two statistical approaches–EdgeR[26] and DEseq[27]–are readily available as packages within the Bioconductor suite. Both reject the *RPKM* normalisation method and propose alternative scaling factors for this purpose.

The EdgeR software uses a normalisation method called *TMM* (trimmed mean of *M*-values), where *M*, true to the tradition of microarray terminology, refers to the fold change in expression between two conditions. The method requires one sample to be the reference, which is used as the basis for calculating the normalising *TMM* factor. The *TMM* factor is then used to adjust the library size of a sample. The normalised RC_N is calculated by dividing the raw *RC* by the adjusted library size. TMM is defined as a weighted mean of log ratios of the raw *RC* between the sample and the pre-defined reference. The weight is calculated to account for the fact that the variance is lower for genes with higher *RC* values. If most genes are not differentially expressed–an assumption made by this approach–the factor should be close to 1. If not, it provides a value, which can be used to adjust the library size. In computing the TMM, the program ignores genes with high RC or *M*-values. This is done to account for the possibility that some genes can be expressed at high levels in condition *A*, but not in *B*; though all the other genes are expressed at similar levels between *A* and *B*, the limitation on the total sequencing 'real estate' could make these show up as less expressed in *A* relative to *B*. By ignoring outliers in calculating the normalisation factor, EdgeR circumvents this confounding factor.

The DEseq package also assumes that most genes are not differentially expressed. However, it differs from EdgeR in that it does not require a reference sample. The scaling factor that DEseq calculates is the ratio of the median of *RC* values across all genes in a sample, and its geometric mean across all samples. Normalised RC_N values are obtained by dividing the raw *RC* by the above correction factor for that sample.

---

25 Dillies M. A., Rau A., Aubert J., Hennequet-Antier C., Jeanmougin M., Servant N., Keime C., Marot G., Castel D., Estelle J., Guernec G., Jagla B., Jouneau L., Laloë D., Le Gall C., Schaëffer B., Le Crom S., Guedj M. and Jaffrézic F. French StatOmique Consortium. 2012. 'A comprehensive evaluation of normalization methods for Illumina highthroughput RNA sequencing data analysis.' *Briefings in Bioinformatics.*

26 Robinson and Oshlack. 2010. 'A scaling normalization method for differential expression analysis of RNA-seq data.' *Genome Biology* 11: R25.

27 Anders and Huber. 2010. 'Differential expression analysis for sequence count data.' *Genome Biology* 11: R106.

Other methods use a descriptor of the distribution of *RC* values across all genes from that sample as the denominator in normalisation. This number could be the upper quartile or the median of the distribution of *RC* values. In our lab, we used the mode of the distribution as the denominator, a normalisation method which appears to have worked across the set of RNA-seq data we had generated, providing good correlations with fold change measures obtained from EdgeR; however, this requires more extensive testing. Some methods invoke the quantile normalisation method, adopted by the RMA procedure in normalising microarray data, for RNA-seq data. However, the fact that microarray intensities follow a continous distribution whereas *RC* values form a discrete distribution mean that the best performing statistical methods (namely EdgeR and DE-seq) are those which have been developed from scratch considering the peculiar characterisitcs of RNA-seq data. This consideration holds true even more for statistical methods used to decide differential expressions, which are described below.

5.4.2.2 Measuring differential expression from read count data

We had briefly introduced the problem of testing whether a gene is differentially present/absent between two samples, in the context of comparative genome hybridisation using microarrays. The problem statement is the same for measuring gene expression differences using next-generation sequencing data (or for that matter, microarrays). To quote Anders and Huber from their publication describing the DEseq package, 'we would like to use statistical testing to decide whether, for a given gene, an observed difference in read counts is significant, that is, whether it is greater than would be expected just due to natural random variation'.

Let us consider a gene *i*, whose *RC* values from replicate experiments are $RC_{i,j\varepsilon A}$ and $RC_{i,j\varepsilon B}$ for conditions *A* and *B* respectively, where *j* represents the various replicate samples for that condition. If the corresponding normalised read counts are $RC^N_{i,j\varepsilon A}$ and $RC^N_{i,j\varepsilon B}$ and the mean over the replicates are $\overline{RC}^N_{i,A}$ and $\overline{RC}^N_{i,B}$, then the objective is to test the null hypothesis $\overline{RC}^N_{i,A} = \overline{RC}^N_{i,B}$. To achieve this, a suitable probability distribution representing the spread of the data across replicates has to be defined. Based on this distribution, the null hypothesis can be evaluated.

A simple distribution that has been used to model count data is the Poisson distribution–also used to calculate the differential expression. However, its properties are restrictive in the sense that by definition, the mean (μ) is equal to the variance (σ^2). This works very well for technical replicates. However, data from biological replicates display what is known as 'over dispersion', where variance is greater than the mean. An alternative distribution, which has been used to account for this problem is the negative binomial distribution,[28] where the variance is not less than the mean. Now, the read count data can be modelled as

[28] For the negative binomial distribution, $P(X = x) = {}^{x+r-1}C_{r-1}\, p^r (1-p)^x$; $p = \mu/\sigma^2$ and $r = (\mu^2)/(\sigma^2 - \mu)$.

a negative binomial distribution $RC_{i,j} = NB(\mu_{i,j}, \sigma^2_{i,j})$. Then, the test for differential expression statistically evaluates whether the mean of the negative binomial distribution representing read counts across replicates for condition *A* differs from that for condition *B*.

As previously mentioned for microarray data, the number of replicates is typically small and therefore, estimating both $\mu_{i,j}$ and $\sigma^2_{i,j}$ reliably is difficult. However, the programs we have visited in the previous section, EdgeR and DEseq, have developed procedures where the power of data obtained across the entire set of genes is used to estimate per gene variances. This remains a field of active research. A third method called baySeq also uses the negative binomial distribution to model the data. A survey of methods for computing differential expression ranks baySeq above all other methods, performing slightly better than edgeR and DEseq.[29] Another method called TSPM[30] states that the over dispersion problem may not be applicable to all genes. The method therefore tests, for each gene, whether it shows over dispersion, and depending on the result, uses variants of the Poisson distribution to model the data. The above-mentioned survey states that TSPM performs well only when the number of replicates is high; otherwise, it falls way behind the negative binomial-based methods.

To summarise, analysis procedures for computing and comparing gene expression from RNA-seq data are in general different from those for microarray data. This is primarily because of the differences between the two approaches in the characteristics of the data they produce. These methods are under continuous evaluation and development, and this remains an important area of research.

5.5 Gene expression at high temporal resolution using fluorescent reporters

It is undeniable that DNA microarrays and next-generation sequencing approaches such as RNA-seq have enabled unprecedented large-scale surveys of gene expression patterns, as well as experimental annotation of genomes for many organisms. However, these come with the disadvantage that sample preparation requires invasive and labourious techniques, including cell lysis, RNA stabilisation and purification. Since transcriptional changes can be effected in a matter of minutes, there is much interest in describing the transcriptional state of cells at high temporal resolution in dynamic environments. Even the routinely used batch culture system is a highly dynamic setting wherein the composition of the medium changes continuously, with concomitant changes in the growth

29 Kvam V. M., Liu P. and Si Y. 2012. 'A comparison of statistical methods for detecting differentially expressed genes from RNA-seq data.' *Americal Journal of Botany* 99: 248–56.

30 Auer and Doerge. 2011. 'A two-stage Poisson model for testing RNA-seq data.' *Statistical Applications in Genetics and Molecular Biology* 10: 26.

rate and the physiology of the cells. Next, performing gene expression assays using microarrays and RNA-seq provides an average picture of the population. However, there could be considerable cell-to-cell variability in gene expression levels of many genes. Single-cell transcriptomics of bacteria with RNA-seq is still in its infancy.[31] Thus, it is safe to say that analysing gene expression at a single-cell level and/or at a high temporal resolution is not easily achieved using microarrays and RNA-seq.

A solution to these problems is the use of fluorescent reporters (Fig. 5.11). The principle is to clone a gene for a fluorescent protein such as the green fluorescent protein (GFP) downstream of a promoter of interest. This is easily done using plasmids. The fluorescent protein should display a high level of stability and be

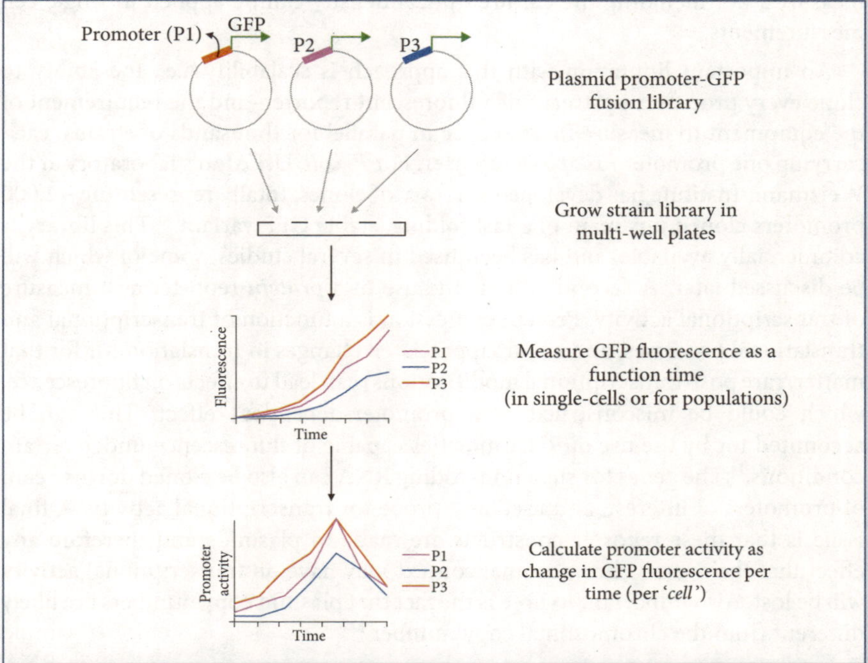

Figure 5.11 *Fluorescence for promoter activity.* This figure shows a schematic approach for measuring transcriptional activity of promoters in live cells, at high temporal resolution, using fluorescence reporters. These reporters can be used to measure fluorescence from a population of cells, or from single cells. A library of promoter-GFP fusions for *E. coli* has been developed by Uri Alon's laboratory.

31 Kang Y., Norris M. H., Zarzycki-Siek J., Nierman W. C., Donachie S. P. and Hoang T. T. 2011. 'Transcript amplification from single bacterium for transcriptome analysis.' *Genome Res.* Jun; 21, 6: 925–35.

able to fold into a fully functional form rapidly. The promoter when activated will lead to the protein being expressed and accumulated in the cell. The fluorescence emitted by the protein thus produced can be quantified from live cell populations using simple fluorimeters without the need for invasive techniques. Fluorescence from single cells can be measured using fluorescence microscopes or flow cytometers. Since the fluorescent protein is expected to be stable with minimal degradation, and also display minimal lag time between synthesis and emitting fluorescence, any increase in fluorecence between two adjacent time-points gives a measure of the production of the reporter protein over that time window. This is in turn a measure of the activity of the promoter to which the reporter is fused. The promoter activity is thus given by $(d(\mathrm{GFP})/dt)/(\mathrm{OD})$, where GFP is the fluorescence, t is time and OD is the optical density of the culture. Similar measures, not including the culture optical density can be applied to single cell measurements.

An important limitation with this approach is scalability–i.e., the ability to clone every promoter upstream of a fluorescent reporter–and the requirement of the equipment to measure fluorescence in parallel for thousands of strains, each carrying one promoter fused to a reporter. For *E. coli*, Uri Alon's laboratory at the Weizmann Institute has developed a library of clones, totally representing ~2,000 promoters cloned upstream of a fast folding, stable GFP variant.[32] This library is commercially available, and has been used in several studies, some of which will be discussed later. A second issue is the use of a *protein* reporter as a measure of transcriptional activity. Protein expression is a function of transcriptional and translational processes, and any unappreciated changes in translation (or for that matter, rare post-transcriptional modifications) can lead to effects on fluorescence, which could be misconstrued as a promoter-dependent effect. This can be accounted for by the use of RNA moieties capable of fluorescence under certain conditions.[33] The genes for such non-coding RNA can also be cloned downstream of promoters of interest, and used as a probe for transcriptional activity. A final issue is that these reporter constructs are made in plasmids and therefore any effect that the overall chromosomal context may have on transcriptional activity will be lost. Also important to note is the fact that plasmid copy numbers are likely different from the chromosomal copy number.[34]

---

[32] Zaslaver A1., Bren A., Ronen M., Itzkovitz S., Kikoin I., Shavit S., Liebermeister W., Surette M. G. and Alon U. 2006. 'A comprehensive library of fluorescent transcriptional reporters for *Escherichia coli*.' *Nature Methods* 3: 623–28.

[33] Paige J. S., Wu K. Y. and Jaffrey S. R. 2011. 'RNA mimics of green fluorescent protein.' *Science* Jul 29; 333(6042): 642–46.

[34] More recently, chromosomal fluorescent fusion libraries have been constructed on a large-scale. See: Taniguchi Y., Choi P. J., Li G. W., Chen H., Babu M., Hearn J., Emili A. and Xie X. S. 2010. 'Quantifying *E. coli* proteome and transcriptome with single-molecule sensitivity in single cells.' *Science* Jul 30; 329, 5991: 533–38.

5.6 Constructing transcriptional regulatory networks: ChIP-chip and ChIP-seq

As described earlier, a variety of DNA binding proteins play vital roles in performing and regulating processes involved in gene expression. Obviously, RNA polymerase itself is a DNA binding protein complex. Many transcription factors bind to specific sites on the DNA and activate or repress transcription. RNA-seq and microarray-based measures of gene expression provide abundance values for RNA molecules in a cell. However, the extent to which these measures describe transcription itself is not clear. This is because the levels of an RNA molecule in a cell is a function of not only its synthesis, but also its degradation. Even within the process of transcription, differences between two RNA molecules in their levels might arise from differences in rates of transcription initiation, or elongation. These differences are difficult to capture using RNA-seq. However, if one could measure *in-vivo*, and on a genome-wide scale, the occupancy of RNA polymerase across genes, and also quantify differences in occupancy between the promoter and the gene body, then the contribution of transcription to gene expression levels could be quantified. Such a technique should also enable us to identify the DNA binding sites of various transcription factors under different conditions, and thus catalog the condition-specific regulatory targets of these transcription factors. These measurements, in conjunction with the appropriate gene expression measurements, either in the form of RNA-seq or as RNA polymerase occupancy map of a genome, enables us to describe the direct and indirect consequences of transcription factor–DNA binding interactions to gene expression. Such information serves as a first step towards understanding the functions of a transcription factor. This, when available for a large number of transcription factors in an organism, has permitted large-scale network analyses, which attempt to discover general patterns in the architecture and evolution of the transcriptional regulatory apparatus.[35]

Now, how does one obtain such protein–DNA interaction data? For a few model organisms, decades of molecular work have painstakingly characterised the properties of specific interactions. The list of such documented interactions for *E. coli* has grown to such an extent as to provide the content for a database called RegulonDB (see below). However, the genomic approaches, which have emerged in the last 10 years or so, allow us to multiply our knowledge several tens of times in the matter of a single research study. Our ability to derive biologically meaningful information from this new-found tool to accumulate such masses of data is a different matter altogether, and we believe that the example studies cited here will provide some inspiration.

---

[35] Chalancon and Madan Babu. 2011. 'Structure and Evolution of Transcriptional Regulatory Networks.' *Bacterial Stress Responses*. Edited by Storz and Hengge. 2nd Edition. ASM Press, USA.

The class of methods that allow genome-scale identification of protein–DNA interactions includes ChIP-chip and ChIP-seq (Fig. 5.12).[36] Here 'ChIP' refers to chromatin immuno-precipitation, and '-chip' and '-seq' refer to microarrays and next-generation sequencing respectively. The ChIP is the experimental procedure, whereas the '-chip' and the '-seq' are the techniques that provide the readout. These are merely a genome-scale extension of the molecular biology technique–ChIP-qPCR, where the binding of a protein to a pre-determined DNA site is measured using quantitative PCR. ChIP is a biochemical method in which

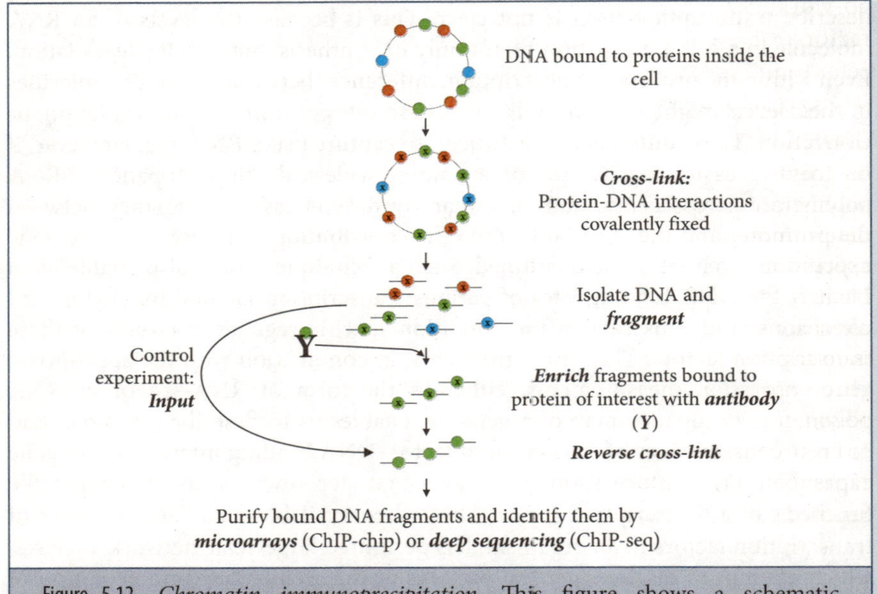

Figure 5.12 *Chromatin immunoprecipitation.* This figure shows a schematic representation of a ChIP-chip/-seq procedure for a protein of interest (green circle). The input control experiment is marked. For a mock-IP experiment, a dummy antibody is used in place of one that is specific to the protein of interest.

protein–DNA interactions are 'frozen' by covalent crosslinking inside live cells,[37] using chemicals such as formaldehyde. Since this is performed on live cultures, the interactions thus captured reflect *in-vivo* events. Then cells are lysed, and the DNA–protein complex is fragmented randomly. Next, fragments associated with the protein of interest are isolated using an antibody against the protein. The crosslinks are reversed and the bound DNA fragments are identified either

[36] Park. 2009. 'ChIP-seq: Advantages and challenges of a maturing technology.' *Nature Reviews Genetics* 10: 669–80.

[37] Thus killing them.

by hybridisation to a microarray (-chip) or by next-generation sequencing (-seq). All the advantages of deep sequencing over microarrays, discussed in previous chapters and sections, are applicable here, and ChIP-seq today is clearly preferred over ChIP-chip in most laboratories. The technique is high-throughput in the sense that in theory, all binding sites of a protein can be identified in a single experiment. There are also several techniques, which have been used to identify protein-binding DNA motifs *in-vitro* in a high-throughput fashion. For example, Professor Ishihama's laboratory has performed such an experiment and identified binding motifs of ~300 *E. coli* transcription factors,[38] just as Jussi Taipale and co-workers have done so on a larger scale for ~1,000 human transcription factors.[39] These are not discussed in this book, but the reader is encouraged to refer to the relevant primary literature.

ChIP-chip and ChIP-seq are not magic bullets that immediately provide a list of DNA binding sites of a protein. Various experimental factors affect the quality of these experiments, which necessitate considerable–if not algorithmically sophisticated–data analysis procedures. The first experimental factor that determines ChIP quality is the antibody. The antibody must be sensitive and specific enough to give a high level of enrichment of bound DNA over the background; and needless to say, not cross-react with unintended targets. An alternative to antibodies generated in a custom fashion to a protein of interest is to tag the gene with the DNA sequence of a standard epitope. Genes can be chromosomally tagged with epitopes relatively easily in model organisms such as *E. coli*, for which tools for genetic manipulation are well-established. It is crucial to ensure that such tags do not interfere with the function of the protein. One such tag is the FLAG epitope for which ChIP-grade antibodies can be purchased from reputed catalogues.

A ChIP experiment is not entirely free of systematic artifacts. There could be biases in fragmentation wherein the more open regions of the DNA fragment more easily than those protected by certain DNA-binding proteins. Certain DNA regions may be more amenable to error-free sequencing, in a sequence or structure-dependent manner. Therefore, it is important to compare any signal from a ChIP-seq/ChIP-chip experiment with that from the same region in a suitable control experiment. One such control experiment is the mock-IP in which the IP is performed without an antibody. This is as close as it gets to a real ChIP experiment. However, it produces inconsistent data as very little material gets pulled down in the absence of the antibody. A second type of control is called

38 Ishihama. 2010. 'Prokaryotic genome regulation: Multifactor promoters, multitarget regulators and hierarchic networks.' *FEMS Microbiology Reviews* 34: 628–45.

39 Jolma A., Yan J., Whitington T., Toivonen J., Nitta K. R., Rastas P., Morgunova E., Enge M., Taipale M., Wei G., Palin K., Vaquerizas J. M., Vincentelli R., Luscombe N. M., Hughes T. R., Lemaire P., Ukkonen E., Kivioja T. and Taipale J. 2013. 'DNA-binding specificities of human transcription factors.' *Cell* 152: 327–39.

the input, which is fragmented DNA prior to the IP. This corrects any biases created during fragmentation or amplification. It is to be noted that there is no consensus on what the most appropriate control experiment is.

The primary goal of a ChIP-seq experiment is to identify binding regions for a protein on the genome.[40] In terms of data analysis, this translates to identifying genomic regions with significant enrichment in the number of reads mapped. Such regions are generally referred to as *peaks*, and software that identify peaks from ChIP-seq data are called *peak callers*. The first step in any peak calling pipeline is the mapping of reads to the reference genome. Then, the genome may be divided into small windows and the number of reads mapping to each window calculated. Since the reads represent only the 5′-end of the fragments produced by the experiment, the mapped position is shifted by 50% of the average fragment length so that it is more likely to represent the mid-point of the fragment; or each read may be extended to the average fragment length. Better estimates of the fragment length can be obtained using paired-end sequencing data. The coverage within each window is assessed to test whether it represents an 'enrichment' based on several criteria. One criterion is a threshold for read coverage, which may be defined based on assumptions for the background read coverage distribution. A second criterion would be a significant increase in read coverage when compared with the same region in a control experiment. More recent peak calling methods exploit the directionality of ChIP-seq data to more precisely define peaks than possible otherwise. Since only the 5′-ends of fragments are sequenced, the actual binding site should be flanked on either end by peaks determined solely from the plus-strand-mapped reads and those from the minus-strand-mapped reads. A combined distribution, which defines binding sites with high precision, can then be constructed (Fig. 5.13). A recent survey of many peak-calling methods showed that these were largely comparable across many parameters, with a discriminatory parameter being the precision in pinning down the exact location of the binding site.[41]

The second goal of ChIP-seq experiments is differential binding of the same protein to the genome under different conditions, or a comparison of the binding of orthologous DNA–binding proteins to their respective genomes. Such analysis can be pursued at several levels. The simplest approach determines whether a peak called in one condition is also observed in the second. The value of this approach depends on the read coverage thresholds that are used. Threshold definition involves subjective calls, and a peak identified in one sample may fall just below the threshold in another. Further, there could be significant quantitative

---

40 Furey. 2012. 'ChIP-seq and beyond: New and improved methodologies to detect and characterize protein–DNA interactions.' *Nature Reviews Genetics* 13: 840–52.

41 Wilbanks and Facciotti. 2010. 'Evaluation of algorithm performance in ChIP-seq peak detection.' *PLoS One* 5: e11471.

difference between samples in the binding signal for a peak, which is called in both samples. Thus, the ability to perform quantitative comparisons between ChIP-seq profiles assumes importance. For this purpose, it is possible to adopt methods for performing differential expression analysis with RNA-seq data (such as DEseq and EdgeR). Dedicated tools such as MAnorm[42] have been developed for ChIP-seq data. These consider the possibility that there could be systematic, non-biological differences between samples in their signal-to-noise ratios. In

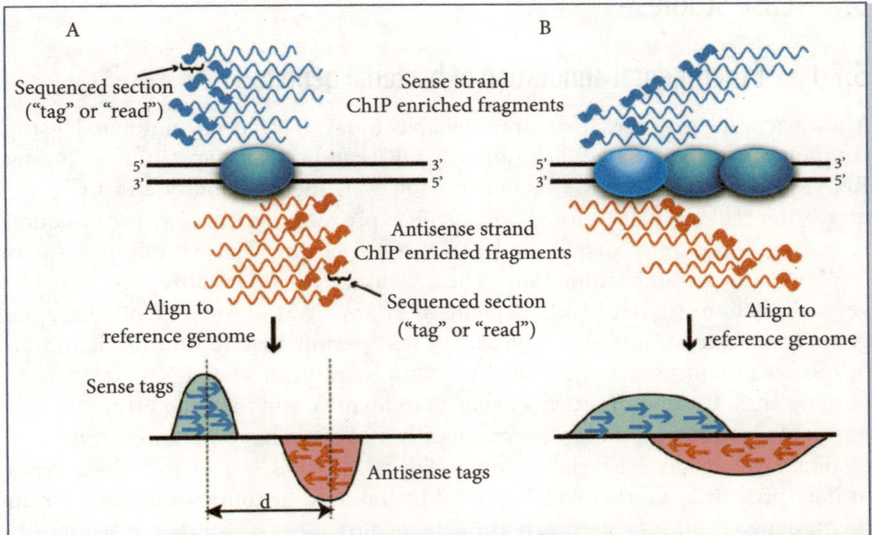

Figure 5.13 *Directionality in read alignments in ChIP-seq data.* The shaded blue oval represents the protein of interest bound to DNA (solid black lines). Wavy lines represent either sense (blue) or antisense (red) DNA fragments from ChIP enrichment. The thicker portion of the line indicates regions sequenced by short read sequencing technologies. Sequenced tags are aligned to a reference genome and projected onto a chromosomal coordinate (red and blue arrows). (A) Sequence-specific binding events (e.g., transcription factors) resulting in sharp peaks (B) Distributed binding events (e.g., histones or RNA polymerase) produce a broader pattern of tag enrichment that results in a less defined bimodal pattern. The figure is reproduced, and the figure legend modified, under the Creative Commons Attribution License from Wilbanks E. G. and Facciotti M. T. 2010. 'Evaluation of Algorithm Performance in ChIP-seq Peak Detection.' *PLoS ONE* 5, 7: e11471. © Wilbanks and Facciotti. 2010.

[42] Shao Z., Zhang Y., Yuan G. C., Orkin S. H. and Waxman D. J. 2012. 'MAnorm: A robust model for quantitative comparison of ChIP-seq data sets.' *Genome Biology* 13: R16.

the process, they make the assumption that most peaks that are common to two samples have similar signals, and cannot necessarily be used to compare conditions leading to global differences in binding signals. The biophysical meaning of ChIP-seq signals,[43] and robust methods for quantitative comparisons of ChIP-seq profiles, remain active areas of research.

5.7 Case studies

5.7.1 Experimental annotation of bacterial genomes

Most bacterial genomes that are available today have been annotated using computational tools. These help identify ORFs and certain types of non-coding RNA encoded in a genome sequence. However, the availability and the ever-increasing reliability of genome-wide gene expression measurement techniques, including tiling microarrays and RNA-seq, permit what is referred to as 'experimental genome annotation'. These techniques help identify regions of the genome that are transcribed, independent of any prior knowledge of where the genes are located. Alongside approaches that permit base-resolution definition of transcript end points, and ChIP-seq/chip identification of RNA polymerase binding sites, these approaches enable us to identify and describe promoter and terminator structures, and in effect, identify and describe the 'transcription unit architecture' of any bacterial genome (Fig. 5.14). It is hoped that these types of data provide resources which can add value to a genome sequence, help in the design of more focused experiments, or provide a substrate for large-scale computational studies integrating multiple data sources. In this section, we will describe a few such resources developed for a few bacterial genomes in recent years.

5.7.1.1 RNA-seq-based annotation of the genome of *Bacillus anthracis*

The first (published) deep sequencing-based characterisation of a bacterial transcriptome was performed for the gram positive pathogen *Bacillus anthracis* by Passalacqua and co-workers.[44] The objective of this work was to annotate the expressed regions of this bacterial genome, and to study the structures of its transcripts with respect to their start sites and transcription units. To maximise the diversity of transcripts captured, which is essential for an experimental

---

43 He X., Chen C. C., Hong F., Fang F., Sinha S., Ng H. H. and Zhong S. 2009. 'A biophysical model for analysis of transcription factor interaction and binding site arrangement from genome-wide binding data.' *PLoS One*. 4: e8155.

44 Passalacqua K. D., Varadarajan A., Ondov B. D., Okou D. T., Zwick M. E. and Bergman N. H. 2009. 'Structure and complexity of a bacterial transcriptome.' *Journal of Bacteriology* 191: 3203–3211.

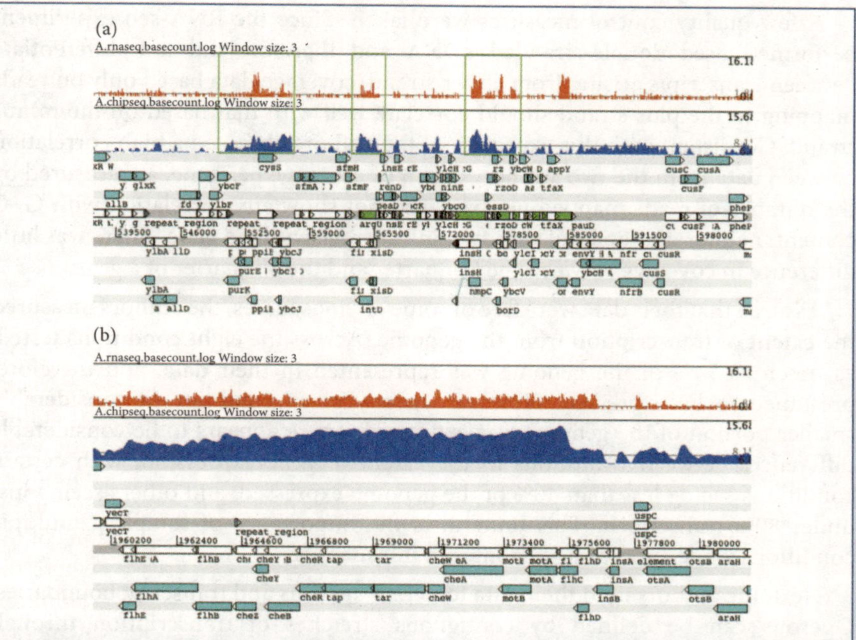

Figure 5.14 *Gene expression and RNA polymerase occupancy.* This figure shows RNA-seq (red) and RNA polymerase ChIP-seq (blue) tracks for regions of the *E. coli* genome. (a) depicts a general correlation (regions marked in green boxes) between the two measures of gene expression over ~60 kb of the genome; (b) shows a close-up of the agreement between RNA-seq and ChIP-seq for the chemotaxis operon. Such data covering multiple growth conditions for the same bacterium – with various important embellishments such as single base-resolution definition of transcript boundaries as developed by the Vogel laboratory – form part of experimental genome annotation projects, some of which are described in this chapter. These data were generated as part of the study published in Kahramanoglou et al. 2011. This figure was drawn using the Artemis Genome Viewer.

annotation project of this nature, the authors sampled several environments, encompassing different phases of growth in two atmospheric conditions. In total, eight conditions were sampled, and four biological replicate experiments performed for each. Since over 90% of the total RNA pool could be ribosomal RNA, 16S and 23S RNA were removed prior to cDNA synthesis. cDNA generated from each RNA pool was subjected to deep sequencing on a SOLiD sequencing platform; a few samples were sequenced on an Illumina system. All sequence reads were mapped back to the reference genome, allowing up to four mismatches for Illumina and five for SOLiD data. Reads which mapped to a single locus were used to generate coverage data, i.e., the number of reads mapping to each base on the genome.

A few quality control measures were taken. Since the RNA-seq experiment performed used double-stranded cDNA and therefore did not differentiate between transcripts arising from either strand, coverage data based only on reads mapping to the plus strand should correlate well with that based on the minus strand. Consistent with this expectation, the authors observed a high correlation between data from the two strands. Depth of coverage per gene, as measured by the number of reads mapped per base, did not show any correlation with G+C content; neither did the local variation in coverage within a gene. There was little difference in coverage between the 5′-quarter and the 3′-quarter of a gene.

Assured that their data were free of some obvious biases, the authors measured the extent of transcription from the genome. Across the eight conditions tested, as much as 94% of the genome was represented in their data, and therefore, presumed to be expressed. However, in any given condition, a considerably smaller portion of the genome was expressed. There appears to be considerable differences between conditions in the extent of gene expression, with certain conditions seeing less than 40% of the genome expressed, and others seeing just under 80% transcribed. This underlines the importance of sampling multiple conditions while annotating genomes in this manner.

Next, the authors used their data to define operons and transcript boundaries. Operons can be defined by contiguous stretches of transcription through intergenic regions. On the other hand, transcript boundaries can be observed as sharp transitions in coverage. Using simple and intuitive rules to define operons and their boundaries, the authors identified ~3,100 transcription start sites (TSS). A TSS is defined as the point upstream of an expressed gene at which coverage first drops to zero. Of these TSS, ~1,300 could be consistently identified across all eight samples. That these identifications were reliable was established using 5′-RACE, a molecular technique for identifying 5′-ends of transcripts at a single-base resolution. The authors identified a few cases where the TSS was present inside previously annotated ORFs. These were interpreted as misannotations, where the start codon had been wrongly identified by computational genome annotation pipelines. The authors also validated their operon predictions using endpoint RT-PCR. They also performed analysis of gene expression levels across the eight conditions, but we will not discuss their results here.

In a more recent study, the same group of researchers performed a strand-aware transcriptome study of *B. anthracis*[45] using previously published protocols[46]

---

[45] Passalacqua K. D., Varadarajan A., Ondov B. D., Okou D. T., Zwick M. E. and Bergman N. H. 2012. 'Strand-specific RNA-seq reveals ordered patterns of sense and antisense transcription in Bacillus anthracis.' *PLoS One* 7: e43350.

[46] Perkins T. T., Kingsley R. A., Fookes M. C., Gardner P. P., James K. D., Yu L., Assefa S. A., He M., Croucher N. J., Pickard D. J., Maskell D. J., Parkhill J., Choudhary J., Thomson N. R. and Dougan G. 2009. 'A strand-specific RNA-seq analysis of the transcriptome of the typhoid bacillus Salmonella typhi.' *PLoS Genetics* 5: e1000569.

as a template. In obvious contrast to their previous strand-unaware RNA-seq experiments, coverage data from the two strands did not correlate at all. Transcripts could now be classified as either sense or antisense depending on whether it emerged from the same or the opposite strand to previously annotated genes. The authors observed widespread antisense transcription. Based on their observation that antisense transcripts could not be detected for ~30% of genes expressed in the sense orientation, the authors suggested that the antisense transcription they observed is not a by-product of sense transcription. As part of an analysis of patterns of antisense transcription, the authors note that the expression levels of antisense transcripts, measured by coverage per base, was much less than those for sense transcription. Analysis of many bacterial genomes had showed that more genes are encoded on the leading strand of replication, than on the lagging strand. In *B. anthracis*, the authors observed greater sense transcription from the leading strand than from the lagging strand; however, antisense transcription appeared to be more prominent on the lagging strand. The authors interpreted this as suggesting that antisense transcription arises from DNA polymerase–RNA polymerase conflicts.

5.7.1.2 The complex transcriptome of a genome-reduced bacterium

For humans and model eukaryotes such as *Drosophila melanogaster* and *Caenorhabditis elegans*, large, consortia-driven projects such as ENCODE[47] and modENCODE[48] have generated genome-scale transcriptional and protein–DNA interaction data, which enable a thorough functional annotation of these genomes. Irrespective of the popular criticism of the manner in which some of these data have been interpreted,[49] these efforts have produced a valuable resource. Along similar lines, a group of researchers from the European Molecular Biology Laboratory (Heidelberg, Germany) and the Centre for Genomic Regulation (Barcelona, Spain) embarked on a project to provide a genome-scale experimental annotation of the genome of one of the smallest

---

[47] ENCODE Project Consortium, Bernstein B. E., Birney E., Dunham I., Green E. D., Gunter C., Snyder M. (+594 collaborators). 2012. 'The ENCODE Project Consortium. An integrated encyclopedia of DNA elements in the human genome.' *Nature* 489: 57–74.

[48] For example: Nègre N., Brown C. D., Ma L., Bristow C. A., Miller S. W., Wagner U., Kheradpour P., Eaton M. L., Loriaux P., Sealfon R., Li Z., Ishii H., Spokony R. F., Chen J., Hwang L., Cheng C., Auburn R. P., Davis M. B., Domanus M., Shah P. K., Morrison C. A., Zieba J., Suchy S., Senderowicz L., Victorsen A., Bild N. A., Grundstad A. J., Hanley D., MacAlpine D. M., Mannervik M., Venken K., Bellen H., White R., Gerstein M., Russell S., Grossman R. L., Ren B., Posakony J. W., Kellis M. and White K. P. 2011. 'A cis-regulatory map of the Drosophia genome.' *Nature* 527–31.

[49] For example, Doolittle. 2013. 'Is junk DNA bunk? A critique of ENCODE.' *Proceedings of the National Academy of Sciences USA*. 110: 5294–5300.

free-living organisms known, the bacterium *Mycoplasma pneumoniae*.[50] The project comprised diverse experiments, including the identification of protein complexes[51] and characterisation of the metabolic network[52] of this organism. What we will discuss here is a set of genomic experiments that these researchers performed to describe the transcriptome of this minimal bacterial genome encoding less than 700 protein-coding genes.

The researchers generated strand-aware RNA-seq data for *M. pneumoniae* grown to a reference time-point in a standard growth medium. In addition, they used tiling microarrays to obtain genome-wide transcriptional profiles at various time-points during a standard growth curve experiment, and following the introduction of stresses such as heat shock and DNA damage. Across these experiments, amounting to over 140 independent tiling microarray or RNA-seq datasets, they observed the expression of all annotated genes. For the tiling microarray data, the researchers used a segmentation algorithm, which identified boundaries separating adjacent regions of distinct expression levels, to define transcripts. In addition to the previously annotated genes, this work identified ~120 novel transcripts, of which ~90 were antisense to known genes. The expression levels of these antisense genes correlated positively with those of their sense counterparts. Genes expressing an antisense transcript were expressed at lower levels than genes without an antisense transcript. This suggested to the authors that these antisense RNA might interfere with the expression of their sense counterparts through double-stranded RNA-mediated mechanisms.

In a manner conceptually similar to the *B. anthracis* study, the *M. pneumoniae* project team experimentally determined operon boundaries. In addition, they also described a staircase pattern of expression within operons, wherein downstream genes were expressed at lower levels than those located upstream within the same operon, suggesting extensive sub-operonic transcriptional polarity. Finally, by combining these tiling microarray data with standard gene expression microarrays

[50] Güell M., van Noort V., Yus E., Chen W. H., Leigh-Bell J., Michalodimitrakis K., Yamada T., Arumugam M., Doerks T., Kühner S., Rode M., Suyama M., Schmidt S., Gavin A. C., Bork P. and Serrano L. 2009. 'Transcriptome complexity in a genome-reduced bacterium.' *Science* 326: 1268–1271.

[51] Kühner S., van Noort V., Betts M. J., Leo-Macias A., Batisse C., Rode M., Yamada T., Maier T., Bader S., Beltran-Alvarez P., Castaño-Diez D., Chen W. H., Devos D., Güell M., Norambuena T., Racke I., Rybin V., Schmidt A., Yus E., Aebersold R., Herrmann R., Böttcher B., Frangakis A. S., Russell R. B., Serrano L., Bork P. and Gavin A. C. 2009. 'Proteome organization in a genome-reduced bacterium.' *Science* 27: 1235–1240.

[52] Yus E., Maier T., Michalodimitrakis K., van Noort V., Yamada T., Chen W. H., Wodke J. A., Güell M., Martínez S., Bourgeois R., Kühner S., Raineri E., Letunic I. P. and Serrano L. 2009. 'Impact of genome reduction on bacterial metabolism and its regulation.' *Science* 27: 1263–1268.

for over 170 conditions, the researchers described sub-operonic transcriptional control wherein certain genes within an operon changed their expression levels in a condition-dependent manner, while other members of the same operon did not.

The researchers stated that the degree of condition-specific gene expression control in this organism, including sub-operonic regulatory mechanisms, cannot be reconciled with the small number of transcription factors predicted from its genome. Thus, the authors further concluded that even bacteria with reduced genomes could have complex transcriptomes.

5.7.1.3 Differentiating between primary and processed transcripts

The studies described above had used conventional or strand-aware RNA-seq/tiling microarrays to annotate transcripts in bacterial genomes. They required a quantitative interpretation of per base coverage data to define transcript boundaries. Therefore, the resolution of these definitions may be debatable. A further caveat to the identification of TSS by conventional RNA-seq is its inability to distinguish between primary and processed transcripts. Other studies have adopted modified deep sequencing approaches to address these issues.

Sharma and colleagues from Jorg Vogel's laboratory described the transcriptome of the pathogen *H. pylori* using what they called 'differential RNA-seq (dRNA-seq)'.[53] Primary transcripts carry a 5′-tri-phosphate (5′-PPP), whereas those arising from the processing of longer transcripts have a 5′-mono-phosphate (5′-P). The dRNA-seq method combines two libraries–one from cDNA generated from the total RNA (–library) and a second from RNA treated with an exonuclease that cleaves RNA with 5′-P but not those with 5′-PPP (+ library). The researchers performed dRNA-seq for *H. pylori* grown in different conditions, including standard media at neutral and acidic pH, in contact with host cells, and in cell culture media. The sequencing was performed using the 454 technology. From these experiments, they identified 5′-ends that were enriched in the + libraries over the corresponding –libraries; in other words, 5′-ends that are not found in the + libraries are likely to be processed and not primary transcript ends. We note here that optimal normalisation methods for these comparisons may still need to be worked out, particularly for higher coverage data generated through short read technologies.

The researchers identified ~1,900 primary TSS across the genome, for ~1,700 annotated genes. Many genes had more than one TSS.[54] A class of TSS were

---

[53] Sharma C. M., Hoffmann S., Darfeuille F., Reignier J., Findeiss S., Sittka A., Chabas S., Reiche K., Hackermüller J., Reinhardt R., Stadler P. F. and Vogel J. 2010. 'The primary transcriptome of the major human pathogen *Helicobacter pylori*.' *Nature* 464: 250–55.

[54] Where two TSS could be identified, the one with the higher coverage is termed primary and the other, secondary.

located in a sense direction inside a previously annotated gene. The researchers could also define TSS for antisense and non-coding RNA. The identification of primary TSS internal to longer operons, showed that sub-operonic transcriptional signatures can be explained by internal transcription intiation events, and not necessarily by the processing of longer primary transcripts. Analysis of promoter sequences around TSS showed the presence of an extended –10 sequence, but no canonical –35 site. The latter appeared to be replaced by a periodic A+T-rich sequence, suggesting that this class of bacteria use a variation of the standard promoter structure to initate transcription.

In addition to describing the 5′-ends of transcripts with high precision, the researchers also identified a plethora of non-coding small RNA[55] in this organism, which is not believed to encode the RNA chaperone Hfq (referred to earlier in this chapter, and discussed in detail later).

In a very recent study, Cortes and colleagues used this method of specifically sequencing 5-′PPP ends to describe the primary transcriptome of the pathogen *Mycobacterium tuberculosis*.[56] An important finding that these researchers made was that over 20% of all expressed genes were in the form of leaderless mRNAs, i.e., there was hardly any 5′-UTR upstream of the translation start site for these genes. The researchers found that many of these genes were particularly over-expressed during starvation. During starvation, the translation machinery is down-regulated, and the ribosome may be maintained in its 70S state, which is not commonly associated with translation initiation.[57] However, some studies have shown that the 70S form of the ribosome can initiate translation at leaderless mRNAs. This allowed Cortes and colleagues to speculate that leaderless mRNAs in *M. tuberculosis* are selected for translation initiation by the 70S ribosome during starvation.

5.7.1.4 The transcriptional architecture of the *Escherichia coli* genome

With experimental genome annotation becoming eminently possible for any bacterium, the model organism *E. coli* was not to be left behind. In fact, a study describing the transcriptional architecture of the *E. coli* genome was one of the earliest (if not the first) such studies to be published. This work, pursued by Cho and colleagues from Bernhard Palsson's laboratory, is unique in that it integrated

---

55 A large number of these RNA molecules were verified by Northern blots. accessed on 10th July, 2014.

56 Cortes T., Schubert O. T., Rose G., Arnvig K. B., Comas I., Aebersold R. and Young D. B. 2013. 'Genome-wide mapping of transcriptional start sites defines an extensive leaderless transcriptome in Mycobacterium tuberculosis.' *Cell Reports* 5: 1121–1131.

57 The initial recognition of the ribosome binding site is performed by the 30S subunit of the ribosome.

multiple types of genome-scale experiments in annotating the organism's transcriptome.[58]

The researchers used four different conditions and obtained gene expression levels using genome tiling microarrays. In addition, they used these microarrays to perform ChIP-chip experiments against the RPO (RNA polymerase). For these ChIP-chip experiments, they used two approaches. In the first, the standard experiment, they found the expected variation in RPO occupancy of the genome across conditions, as different genes are transcribed under different conditions. In the second approach, they inhibited transcription elongation using the antibiotic rifampicin. This forces the RPO to stall at promoters. Moreover, since there is no elongation, there is enough free RPO in any condition to occupy all promoters; therefore, the genome-wide binding profile of RPO in rifampicin-treated cells does not change across conditions. This essentially provides a catalogue of *E. coli* promoters. Integrating the RPO ChIP-chip data with their transcriptome data, the researchers defined ~2,700 transcript segments bounded by RPO binding regions. These segments were termed as 'RPO[59]-guided transcript segments'.

Next, the researchers used a variant of 5′-RACE, followed by deep sequencing to identifiy TSS at a single-base resolution. Unlike the dRNA-seq approach used for *H. pylori*, this approach probably does not differentiate between primary and processed transcripts. Across the four conditions tested, each RPO-guided transcript had–on average–1.6 TSS, suggesting extensive alternative TSS, probably determined by the σ-factor involved in transcription initiation under a given condition. Additional proteomic studies also supported many transcript segments.

Genome annotation had predicted ~2,500 operons in the *E. coli* genome. However, the larger number of TSS identified in this study–along the lines of similar studies described above–suggest a more complex transcriptional structure than previously anticipated.

5.7.2 Bioinformatic analysis of bacterial promoters

Though much of this textbook discusses methods and research into new genomic technologies and their application to understanding bacterial genomes and their expression, this chapter will also discuss bioinformatic studies investigating genome sequences and large compilations of literature-derived molecular data towards understanding aspects of bacterial gene regulatory systems.

---

[58] Cho B. K., Zengler K., Qiu Y., Park Y. S., Knight E. M., Barrett C. L., Gao Y. and Palsson B. O. 2009. 'The transcription unit architecture of the *Escherichia coli* genome.' *Nature Biotechnology* 27: 1043–1049.

[59] Note that the authors use the abbreviation RNAP (and not RPO) for RNA polymerase; therefore, in their paper, these segments are called 'RNAP-guided'.

5.7.2.1 The RegulonDB database and its contribution to gene regulation studies of *E. coli*

Julio Collado-Vides and his colleagues have championed the cause of (quasi) genome-scale bioinformatics for studying gene regulatory networks in *E. coli* and occasionally in other bacteria. Besides publishing several papers analysing the architecture of the transcriptional system of *E. coli*, his laboratory has established a popular database called RegulonDB.[60] This publicly-available database is a compendium of manually-curated data encompassing multiple aspects of the *E. coli* transcriptional machinery. These include data on promoters and σ-factor and transcription factor–gene interactions. Much of these data have been sourced from the primary literature. Being a large one-stop-shop for these types of data, the RegulonDB database provides an opportunity for bioinformaticians to perform various statistical analyses of its data. For the molecular biologist, the database provides the citation for every piece of information it stores, permitting the user to evaluate the quality of the evidence for that data point.

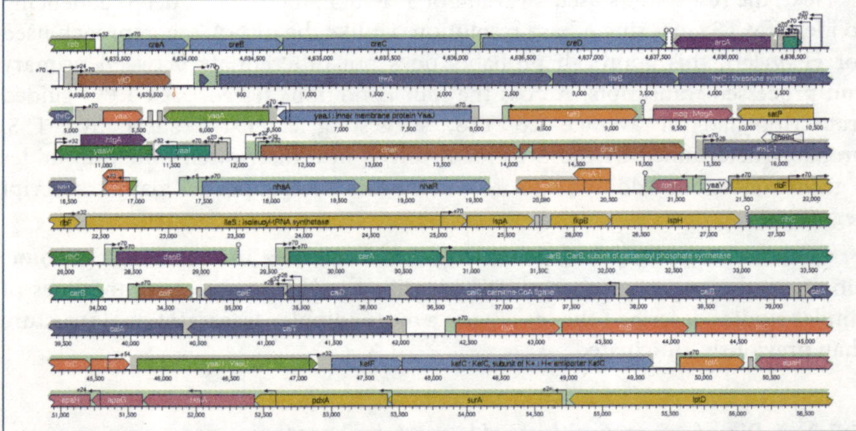

Figure 5.15 *Overview of functional elements of E. coli from Ecocyc and RegulonDB.* An overview of the annotation of a part of the *E. coli* genome, downloaded from the Ecocyc database (http://www.ecocyc.org). Attention is drawn to the annotation of promoters with their respective σ-factors, and to features such as transcription terminators. This regulatory information is provided by the RegulonDB database.

We will revisit RegulonDB when we discuss studies of transcription factor networks in *E. coli*. For the present section on *E. coli* promoters, it suffices to say that RegulonDB stores information assigning promoters and their associated σ-factors to transcription units (Figs 5.4 and 5.15). We will see how these data

60 http://regulondb.ccg.unam.mx/

have been used to extract various statistical features of *E. coli* promoters under the control of the major σ-factor (σD).

5.7.2.2 Features of the *E. coli* promoter

As described earlier in this chapter, the bacterial promoter–with *E. coli* as the paradigm–is often described as bipartite, with a –10 TATAAT sequence and a –35 TTGACA site (Fig. 5.2). The optimal spacing between the two sites is 17 bp. The closer a given sequence is to this structure, the stronger it is as a promoter. However, there is no promoter in *E. coli* that exactly matches this characteristic, i.e., the sequences mentioned above are consensi, reflecting the most common nucleotide at each position. In fact, there is enormous variability in promoter structure across *E. coli* genes. While the –10 element–with some variation in sequence–is present in most σD promoters, the –35 element is dispensable. The extended –10 element, a TG dinucleotide extension at the 5′-end of the –10 sequence, can replace the –35 promoter. Further, minor variations in the spacing between the –10 and the –35 sites can also be tolerated. We had earlier noted the presence of a bent A+T-rich region called the UP element upstream of the –35 element as part of the *E. coli* promoter. Ultimately, it is a *combination* of these features that determine promoter recognition by the RPO and subsequent initiation of transcription. It goes without saying that a description of the structure of promoters is vital to our understanding of transcription. In the face of the variation in structure across promoters, it also becomes important to understand the rationale (if any) behind this variability, and how this variation contributes to the efficiency of transcription initiation. The latter point is difficult to measure on a genomic scale using the *in-vivo* experimental approaches described in this chapter, as any correlation between promoter sequence and transcription initiation requires the analysis to be decoupled from the large number of regulatory factors that influence transcription in the cell; and to the extent of this author's knowledge, *in-vitro* transcription on a genomic scale has not been achieved. Therefore, such studies on a genomic scale have been limited to bioinformatic sequence analysis, supported by data available in databases such as RegulonDB. In this chapter, we will describe a few such studies.

Methods for analysing sequence motifs associated with promoters involve multiple sequence alignments of short sequences upstream of known TSS. This helps identify nucleotides that are found at certain positions more commonly than expected from a background base composition model. These features should then be collated in a scoring system that tests how close a given sequence is to a model promoter. This score might be expected to reflect the strength of the promoter, in the absence of accessory regulators. We do not describe these techniques in any detail, for they are the preserve of classical bioinformatics texts, but present some conclusions arrived at by using such methods.

Starting with a set of nearly 700 experimentally verified TSS (~550 after some filtering steps), obtained from two different databases,[61] Shutzaberger and co-workers determined sequence motifs for the –10 and the –35 sites, which agreed with the consensi obtained earlier using only a few known promoters.[62] They also determined the distribution of the spacing between the two sites, and used these parameters to score a sequence with respect to its promoter-like characteristics. Within the –10 TATAAT consensus, these researchers considered the second A as the zero position and found that the location of this base varied between 8–14 bp upstream of the TSS. They found a very strong conservation of the 3′-most T within this consensus (also reflected in Fig. 5.2, which is based on an analysis performed by the RegulonDB group). This T is likely to be in the minor groove of the DNA. The fact that a single base is specifically conserved in a minor groove[63] suggested to these researchers that the RPO contacts this base in an atypical fashion, which may be in a distorted helix, or through unwound DNA in the open complex, or by the base being flipped out of the helix. As expected from the fact that the –35 element can be replaced by other activator elements, the authors found it difficult to build sequence alignments of this region. Once done, they however noted that the conservation of the TTGACA motif at –35 is prominent in the 5′-end, and dropped sharply towards the 3′-end. This was consistent with the co-crystal structure of a σD homolog with the DNA showing an abrupt termination of contacts after the 5′-region of the –35 element. The remaining space could be occupied by other activating players under the right conditions. The bipartite promoter structure was most apparent at a spacing of 16–18 bp, though a considerable number of sites have sub-optimal spacing ranging from 15 to 20 bp. The bipartite promoter structure with the optimal or sub-optimal spacer lengths explained ~400 of the ~550 TSS studied here. The two promoter elements are best recognised by the RPO when they are on the same face of the double helix, two helical turns apart. However, the two domains of σD that recognise the two sites may rotate relative to each other, or the twist of the DNA may change, thus allowing sub-optimal spacers to be recognised for transcription initiation. The researchers also found instances where the –35 element was effectively replaced by the extended –10 element, in terms of the total conservation score that could be assigned to the promoter. The authors speculate that the extended –10 could be an evolutionary precursor

---

[61] RegulonDB and PromEC databases.

[62] Shultzaberger R. K., Chen Z., Lewis K. A. and Schneider T. D. 2006. 'Anatomy of *Escherichia coli* σ70 promoters.' *Nucleic Acids Research* 35: 771–88.

[63] In B-form DNA, a DNA binding protein cannot specifically distinguish one base within the narrow minor groove; it can only differentiate between an A-T pair and a G-C pair.

to the bipartite promoter. The latter could have evolved to facilitate greater flexibility in gene regulation, for many trans-acting regulators target the −10 or the −35 element. Moreover, the spacing could also allow regulation by changes in DNA geometry/topology such as supercoiling, which can position the two elements in or out of phase with each other.

In an earlier study, Huerta and Collado-Vides performed a computational analysis of promoters upstream of ~600 *E. coli* TSS in the RegulonDB database.[64] They once again found a strong conservation of the −10 element, contrasting with the dispensabilitiy of the −35 site. Furthermore, the absence of the −35 appeared to correlate with the presence of the extended −10. An interesting finding of this study was the discovery that sequences upstream of TSS have many–and not one or two–promoter-like structures. They found as many as 38 promoter-like elements, on average, within 250 bp upstream of the TSS. In fact, many of these sites had a higher score, i.e., greater similarity to the consensus, than the corresponding, experimentally-verified promoter. The researchers note that these promoter-like elements are closely clustered together. Since the footprint of the RNA polymerase is 60 bp, the clustering of multiple promoter-like elements within this stretch might aid promoter recognition and DNA unwinding. An aggregate promoter score across such a cluster might be indicative of the strength of a promoter. These promoter clusters were specific to regulatory regions upstream of promoters, and were rare inside coding regions and in intergenic regions separating convergent genes. Though this could be a direct consequence of elevated A+T-content in gene regulatory regions, the authors argue against this trivial possibility by studying genomes with different A+T-contents and with analysis of shuffled sequences. Whatever the parameter determining the occurrence of promoter clusters, it remains to be understood how the RPO specifically initiates transcription from one (or at best a few) promoters per transcript.

Huerta, Collado-Vides and co-workers–in another study–found that the promoter structure of *E. coli* was conserved in over 40 bacterial genomes across phyla.[65] This is consistent with the conservation of the RPO and the σD. It appears to contrast with the lack of a −35 element, described in the transcriptome study of *H. pylori* by Sharma and colleagues (discussed earlier in this chapter); however this could emerge from differences in the motif searching methods between the two studies. The researchers found however, that the difference between regulatory

---

64 Huerta and Collado-Vides. 2003. 'Sigma70 promoters in *Escherichia coli*: Specific transcription in dense regions of overlapping promoter-like signals.' *Journal of Molecular Biology* 333: 261–78.

65 Huerta A. M., Francino M. P., Morett E. and Collado-Vides J. 2006. 'Selection for unequal densities of σ70 promoter-like signals in different regions of large bacterial genomes.' *PLoS Genetics* 2: e185.

and non-regulatory/coding regions in the occurrence of promoter clusters, seen in *E. coli* and other large bacterial genomes, was absent in the reduced genomes of host-restricted pathogens and symbionts. They argue that this could be explained by reduced purifying selection on reduced genomes, which by itself can lead to the emergence of promoter-like sites by mutation.

5.7.3 DNA topology and its interplay with gene expression

As mentioned earlier in this chapter, though promoter sequence is an important determinant of gene expression, it is relatively static except under evolutionary time-scales where selection might favour a mutant promoter structure that offers a fitness benefit to the population. We had seen an example for this in the *in-vitro* evolution experiment of Richard Lenski where a promoter capture mechanism allowed *E. coli* to utilise citrate as a carbon source in aerobic conditions (see Chapter 4).

'Adaptation' over shorter time-scales occurs by alteration of gene expression typically orchestrated by trans-acting players such as transcription factors. However, DNA structure itself can respond to the environment and regulate gene expression. An example that we have encountered is DNA supercoiling, which changes with cellular conditions, notably, the energy charge. DNA supercoiling can align any sub-optimally spaced promoter elements for effective recognition and transcription initiation by the RPO. We had previously noted that it has been hypothesised that the bipartite promoter structure may have evolved to provide space for regulation by players such as DNA supercoiling.

5.7.3.1 The effect of DNA supercoiling on genome-wide gene expression in *E. coli*

The laboratory of (the now deceased) Nicholas Cozzarelli used microarray-based genomic techniques to investigate the impact of supercoiling on gene expression, and to make interpretations on some aspects of the structure of the *E. coli* chromosome. Peter and colleagues from the laboratories of Nicholas Cozzarelli and Patrick Brown used microarrays to probe the transcriptional response of *E. coli* to loss of negative DNA supercoiling (loss of negative supercoiling is also called DNA relaxation).[66] DNA supercoiling was already well known as a regulator of the expression of selected genes. The transcriptional upregulation of DNA gyrase[67] and the downregulation of topoisomerase I[68] by DNA relaxation establishes a

[66] Peter B. J., Arsuaga J., Breier A. M., Khodursky A. B., Brown P. O. and Cozzarelli N. R. 2004. 'Genomic transcriptional response to loss of chromosomal supercoiling in *Escherichia coli*.' *Genome Biology* 5: R87.

[67] DNA gyrase introduces negative supercoils.

[68] Topoisomerase I relaxes DNA.

homeostatic negative feedback circuit, which should keep the overall superhelical density of the DNA within a certain range. Supercoiling was also known to regulate the expression of selected genes involved in functions not immediately connected to DNA topology, such as amino acid metabolism. The fact that the supercoiling state of the DNA changes with environmental conditions–including changes in osmotic pressure and temperature, besides nutritional state–suggested to the researchers that supercoiling could be a global transcriptional regulator affecting the expression of many genes across multiple cellular conditions. To probe the effects of DNA relaxation on gene expression on a genomic scale, the researchers used a comprehensive experimental approach involving the inhibition of DNA gyrase and topoisomerase IV.[69] Knockouts of these genes are lethal and therefore, partial inhibition by antibiotics or specific point mutations was the way forward. The pleiotropic effects that an inhibitory agent may have on *E. coli*, via routes not necessarily immediately connected to DNA supercoiling, meant that the researchers had to employ multiple inhibitory strategies, all converging on introducing DNA relaxation. Gene expression changes that can be explained consistently across these treatments are likely to result from DNA relaxation and not treatment-specific effects. Thus, the methodology used by these researchers is as follows. They used two antibiotics, one a DNA gyrase inhibitor and the other a topoisomerase IV inhibitor, and found that these quickly relaxed the DNA as determined by superhelical densities[70] measured on a plasmid probe. In addition, they used the same treatments on mutant *E. coli*, which were resistant to the antibiotic, and found that these treatments had little effect on DNA supercoiling in these resistant strains. This serves as a control where the antibiotics are applied, but with little effect on supercoiling; and any other transcriptional effects these antibiotics have can be cancelled out. Next, the researchers used a temperature sensitive DNA gyrase mutant, which functions normally under a 'permissive' temperature, but is defective in an 'inhibitory' temperature. This mutant and the wild type, grown under the permissive and the inhibitory temperatures form a set of samples.

RNA was isolated at various timepoints following the treatment with the antibiotic or shift to the inhibitory temperature, the cDNA synthesised and hybridised to a classical DNA microarray. For each of the conditions tested on the microarray, plasmid supercoiling was also measured. The researchers could bin these samples into two based on whether the DNA was relaxed or not. Now they could perform a differential gene expression analysis between the two groups. Since each group of samples included variability across parameters not connected

[69] Topoisomerase IV relaxes positive supercoils, thus making the superhelical density more negative.

[70] *Superhelical density*: The number of turns added or removed relative to the total number of turns in the relaxed molecule/plasmid, indicating the level of supercoiling (definition from Wikipedia).

to supercoiling, significant changes in expression that are consistent within each bin would likely reflect an effect of the supercoiling state of the DNA. Since the number of samples within each bin was large, they could use a standard t-test to assess the significance of the differential expression. This resulted in a set of ~300 supercoiling-sensitive genes (SSG), representing ~7% of the genes represented on the microarray (Fig. 5.16). In addition to performing a differential expression analysis, the availability of conditions representing a range of supercoiling states allowed the researchers to perform analyses of the correlation between gene expression levels and superhelical densities. A low correlation between the expression level and superhelical density for an SSG determined by the differential

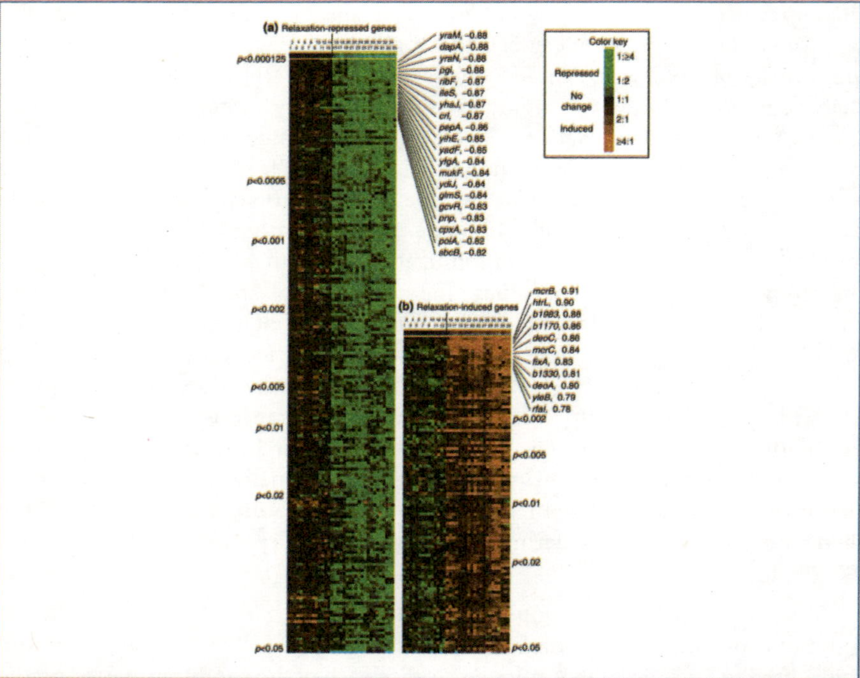

Figure 5.16 *Effect of DNA supercoiling on gene expression in E. coli.* This figure shows a heatmap of gene expression as a function of DNA supercoiling; each row represents a gene and each column a condition in the experiment by Peter et al. 2004; the vertical mark at the top of the heatmap separates conditions where the DNA is relaxed (right) from those where it is supercoiled (left). It was found that relaxation-induced genes generally had a higher A+T-content than an average *E. coli* gene, with the reverse being applicable to relaxation-repressed genes. The figure is reproduced under the Creative Commons Attribution License from Peter, Arsuaga, Brier, Khodursky et al. 2004. 'Genomic transcriptional response to loss of chromosomal supercoiling in *Escherichia coli*.' *Genome Biology* 5: R87. © Peter et al. 2004; licensee Biomed Central.

expression analysis would mean that the SSG only differentiates between two extreme states, and is not sensitive to intermediate changes in supercoiling. However, most SSGs were found to show high correlations. This, in addition to better-resolved time-course analysis, showed that most SSGs respond immediately to DNA relaxation. This short time interval between DNA relaxation and a change in expression levels would further mean that the effect of DNA relaxation on gene expression is largely direct and not mediated by a cascade of factors.

Negative supercoiling should promote DNA unwinding and therefore increase transcription; the opposite being the case for DNA relaxation. Consistent with this expectation, two-thirds of SSGs were repressed by DNA relaxation. However, it was interesting that as many as 100 genes were up-regulated by DNA relaxation. As described earlier, this is unlikely to be an indirect effect. The researchers note that these relaxation induced genes have a higher A+T-content than the average *E. coli* gene. The weaker inter-strand hydrogen bonds expected in A+T-rich DNA would probably favour unwinding even when the DNA is relaxed, and thus promote their expression. In contrast, many genes which were repressed by DNA relaxation had a lower A+T-content, and it may well be the case that efficient DNA melting in these regions requires negative supercoiling.

We note here that another study, performed by Georgi Muskhelishvili's group, investigated the interplay of supercoiling and gene expression in mutants of DNA-binding proteins and found that over ~1,000 genes could be affected by supercoiling across different genetic contexts.[71]

5.7.3.2 Supercoiling-sensitive gene expression as a probe for chromosome conformation

It is well known that in eukaryotes, the large DNA molecule is compacted by nucleosomes alongside other accessory players. The requirement of compaction applies to bacteria as well, since an uncompacted circular chromosome will be ~10^3 times larger than the cell. As is routinely emphasised in textbooks, DNA compaction should retain a structural organisation that is also permissive to critical DNA transactions including DNA replication and transcription, which require unwinding of the highly compacted DNA. DNA supercoiling is a topological aspect of the chromosome, which can compact the DNA as well as promote local unwinding. However, even a single DNA break, which can modulate supercoiling, if allowed to propagate its effects without barriers throughout the chromosome, would be lethal to the cell. Though, various factors, including DNA binding proteins, can act as barriers such that any effect on the topology of the DNA on

71 Blot N., Mavathur R., Geertz M., Travers A. and Muskhelishvili G. 2006. 'Homeostatic regulation of supercoiling sensitivity coordinates transcription of the bacterial genome.' *EMBO Reports* 7: 710–15.

one side of the barrier does not progate to the other. Thus, such barriers create isolated looped domains in the chromosome.

Postow and colleagues from the Cozzarelli lab set out to examine evidence of four different models for the looped domain architecture of the *E. coli* chromosome.[72] The four models differ from each other in (a) whether the location of the barriers is fixed or variable across chromosomes and (b) whether all the domains in a chromosome are of the same size or whether they represent a distribution of sizes. The reader is directed to the introduction of the researchers' paper, which discusses previous literature on this subject.

To answer their questions, the researchers used microarray experiments, alongside the SSGs identified by their lab in the work of Peter and colleagues. Their experimental strategy was as follows. They introduced a restriction enzyme-mediated cut at various points on the chromosome. This should immediately start relaxing the chromosome. The extent to which the relaxation spreads can be measured by calculating the probability of an SSG changing in expression as a function of its distance from the nearest cut–a value that can be measured on a genome-wide scale using microarrays.

The researchers used a restriction enzyme to cut the DNA. A confounding factor in measuring gene expression change immediately after such a treatment is that there are endonucleases which will chew up the DNA from cut ends; as a result, any change in gene expression must be interpreted keeping in mind loss of DNA near such cuts. To be able to do this, the researchers first hybridised genomic DNA after introducing a cut, and performed a comparative genome hybridisation (see Chapter 3) experiment comparing DNA from a cut chromosome and that from wildtype controls. This experiment showed that in the time-scales that the researchers used, only ~3 kb of DNA around a cut was degraded. Next, the researchers hybridised cDNA from the two types of cells to a microarray. Then, for each gene, they plotted fold change in expression between the two cell types as a function of the distance of the gene from the nearest restriction site. Non-SSGs generally seemed to show little change in expression, and the fold change v. distance curve was fairly flat. On the other hand, SSGs showed strong trends. In particular, genes that were repressed by relaxation showed higher fold changes closer to a cut than further away. As the distance between a cut and an SSG increased, up to a limit of ~10–15 kb, the probability that the SSG will respond transcriptionally to a cut decreased. Beyond the 10–15 kb range, the curve flattened.

Next, the researchers asked whether the domain boundaries in the chromosome were fixed across cells. To address this issue, they selected a 200 kb chromosomal region which was devoid of a site for the restriction enzyme used. This region also contained two SSGs. Now the researchers artificially introduced a restriction site at various points within this region, and measured the effect of a cut on the

[72] Postow L., Hardy C. D., Arsuaga J. and Cozzarelli N. R. 2004. 'Topological domain structure of the *Escherichia coli* chromosome.' *Genes and Development* 18: 1766–1779.

expression of the two SSGs. If the domain boundaries were the same across different chromosomal molecules, then one could expect a cut at sites close to SSGs to lead to a change in expression; and a cut that is further away, separated from the SSG by a domain boundary, will have no effect on the SSG's expression. On the other hand, if domain boundaries were variable, then probabilities come into play: Closer the site is to an SSG, less the chance of finding a domain boundary separating the two. Therefore, in this case, one would expect a smooth curve between distance and fold change in the SSG expression. The result of the experiment showed the latter, suggesting that domain boundaries are not fixed, and differ from cell to cell.

Then the researchers asked whether each domain is of the same size or of variable size. To assess this, they performed Monte Carlo simulations, wherein they calculated the probability of finding an SSG and a restriction cut within the same chromosomal domain. Simulations were run for various assumptions: Domain sizes being the same or variable, with the (average) domain size itself being a parameter. Based on their microarray data, the authors could calculate the above probability. Thus, simulations provide an expectation for each domain model, whereas microarray data are the observations. A comparison of the observation with the simulations suggested that domain sizes were not fixed, but followed a distribution, and that the average size of a domain should be around 10 kb. This could be verified by the authors using careful electron microscopy experiments of purified nucleoids.

Thus, using DNA microarrays in a non-standard manner, Postow and colleagues were able to shed light on certain aspects of the topology of the *E. coli* chromosome.

5.7.3.3 Chromosome conformation capture on a genome-wide scale

Recent developments in deep sequencing have catalysed the emergence of a wide range of creative applications addressing different aspects of chromosome biology and gene expression. Among these, methods that lead to the study of the conformation of the chromosome are among the most popular. These methods measure the probability that two regions of the chromosome, distant in the primary sequence, contact each other in space, and calculate this probability for every pair of segments in the chromosome. This measure is generally referred to as a contact frequency or contact probability. A 'standard curve', relating contact frequency to the physical distance between two chromosomal loci in a cell, can then help build 3D models of the chromosome. The starting point for obtaining contact frequencies is the technique called 3C (chromosome conformation capture; Fig. 5.17).[73] This involves cross-linking DNA in the cells, which ensures that

[73] Dekker J., Rippe K., Dekker M. and Kleckner N. 2002. 'Capturing chromosome conformation.' *Science* 295: 1306–1311.

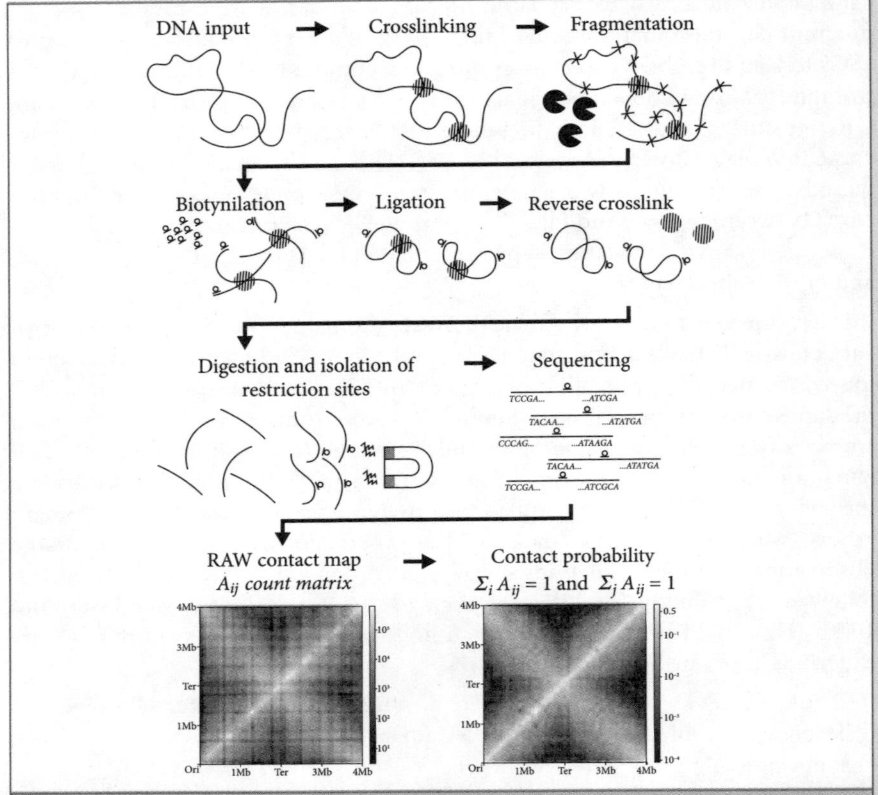

Figure 5.17 *Schematic representation of a chromosome conformation capture protocol.* DNA is cross-linked in the cell so that regions that may be distal in primary sequence but adjacent in space are cross-linked together. A restriction enzyme is used to fragment the cross-linked DNA and the ends of fragments ligated together. This ensures that spatially proximal chromosomal regions are now covalently linked to be adjacent in primary sequence. After reversing the cross-links, the DNA is fragmented and those spanning a ligation site specifically isolated (using a *carbon copy* as in 5C, or by using biotinylated nucleotides during ligation and isolating it using streptavidin as in Hi-C). Paired-end sequencing of DNA on either side of the ligated site helps link loci together and generate contact frequency/probability maps. Figure credit: Vittore Scolari, National Centre for Biological Sciences and University Pierre Marie Curie.

regions of the chromosome that were close together in space get covalently linked, and therefore fixed. Next, the cross-linked DNA is cut with a restriction enzyme, and the ends of fragments ligated together. This makes sure that the cross-linked segment pairs are now adjacent to each other in primary sequence. The DNA produced after reversing the crosslinks is called the 3C library. To reiterate, the

DNA in a 3C library contain sequentially distal, but spatially proximal segments of the chromosome together in the same short sequence. This acts as a substrate for different quantification approaches, including high-throughput deep sequencing and the relatively low throughput qPCR, which measure the contact frequency between two chromosomal segments.

On the bacterial front, genome-wide measurements of chromosome conformation have been performed to high standards for *Caulobacter crescentus*, a model organism for studying asymmetric cell division. The two daughter cells produced after every division event are phenotypically distinct. One is called a stalk cell, which is an adherent cell that can undergo further DNA replication and cell division. The other is called the swarmer cell, which is motile, but does not divide further, until it differentiates into a stalk cell. In a certain strain of *C. crescentus*, newly formed swarmer cells can be separated from their sibling stalk cells and other pre-divisional cells, thus facilitating isolation of a synchronous population of swarmer cells. Genomic information derived from such a synchronous population will be representative of a single cell type, and not an average across cells, which despite being physiologically similar (for example: Mid-exponential phase of growth) could be in different cell divisional stages (pre-replication, post-replication etc.). This could be an important consideration for chromosome conformation studies, as DNA replication and cell division are important determinants of chromosome structure.

Umbarger and colleagues, together representing several laboratories, performed a genome-wide analysis of contact frquencies in a 3C library of chromosomes from a synchronous swarmer cell population of *C. crescentus*[74] (Fig. 5.18). For quantifying contact frquencies, they used a procedure termed as 5C (3C carbon copy). As the name implies, a carbon copy of the sequence surrounding the ligation site in a 3C library is made, and then amplified before being subjected to paired-end deep sequencing. In more detail, single-stranded DNA probes with the following characteristic are used. One part of the probe is complementary to one side of the genomic sequence around each restriction site of interest, and the other is complementary to a universal PCR primer of choice. The 3C library is denatured and then annealed to the set of probes used. Then the probe pair that anneals on either side of a restriction site on the 3C library is ligated together. This represents a carbon copy of the informative part of the 3C library. The carbon copy is now amplified using the universal primer of choice, producing enough DNA for paired-end deep sequencing. If one member of a read pair maps[75] to segment A of the chromosome and the other to segment B, then this qualifies as evidence

---

[74] Umbarger M. A., Toro E., Wright M. A., Porreca G. J., Baù D., Hong S. H., Fero M. J., Zhu L. J., Marti-Renom M. A., McAdams H. H., Shapiro L., Dekker J. and Church G. M. 2011. 'The three-dimensional architecture of a bacterial genome and its alteration by genetic perturbation.' *Molecular Cell* 44: 252–64.

[75] Mapping of sequence read to the reference genome is performed using standard procedures.

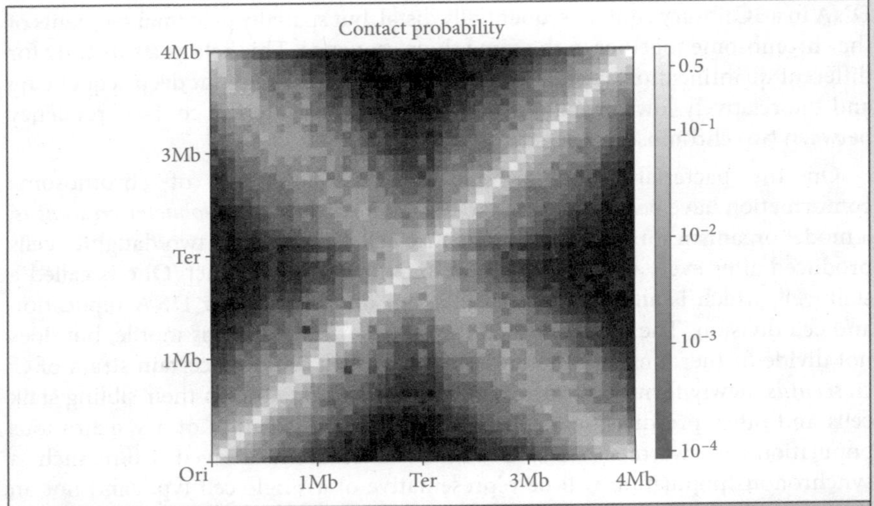

Figure 5.18 *Contact probability map of Caulobacter crescentus genome.* This figure shows the contact probability map of the *C. crescentus* genome, based on data generated by the 5C experiments of Umbarger and colleagues (2011). Chromosomal coordinates are lined up along the horizontal and the vertical axes. The origin and terminus of replication are marked. Black indicates low contact probability and white indicates a high probability. There is low probability of contact between the terminus and the origin, and a high contact probability between regions in register on either side of the origin-terminus axis (marked by the white line along the diagonal running from the top-left to the bottom-right). Figure credit: Vittore Scolari, National Centre for Biological Sciences and University Pierre Marie Curie.

of an interaction between segments A and B. The proportion of read pairs that so links two segments together is a measure of the contact frequency between those two regions. In the *C. crescentus* study, almost 29,000 probe pairs were used, and therefore, interactions between the same number of chromosomal segment pairs could be interrogated. Prior to analysing their 5C data, Umbarger and co-workers first determined whether their 3C libraries reflected known properties of the *C. crescentus* chromosome orientation. It had been previously shown that the origin of replication and the terminus reside on opposite poles of the cell and therefore, rarely interact. By performing qPCR analysis of their 3C libraries, the researchers determined that the frequency of interaction between the origin and the terminus is at least 2^6-times less than that between neighbouring loci within the origin or the terminus regions. Having established that their libraries were reliable, the researchers plotted a contact frequency map of the *C. crescentus* chromosome (Fig. 5.18). In this map, chromosomal loci are arranged in linear order along the *x*- as well as the *y*-axis. The left-and the right-most (equivalent to the bottom and the top-most) ends of the map represent the origin, and the mid-point along either axis, the terminus. Each *x,y* coordinate is coloured according to the contact frequency between the two segments. To reiterate, this frequency is measured by the number of times the two segments were found as a pair in the same paired-end sequencing data. The contact frequency map for *C. crescentus* showed an X-shaped pattern, with the two diagonals intersecting close to the origin and the terminus.[76] The main diagonal represents close-range contacts within the same chromosomal arm, whereas the second diagonal shows in-register contacts between the two chromosomal arms. This indicated that the chromosome was ellipsoidal in shape with extensive contacts between segments on either side of the origin–terminus axis (the two arms of the chromosome), with the origin and terminus themselves being far apart from one another. Using fluorescence reporters, the researchers built a calibration curve, which related contact frequencies from their 5C data to real physical distance between chromosomal segments. Converting all contact frequencies to distances, the authors developed 3D structural models of the chromosome. Several 3D models fit the data, but all belonged to the same class of structures, which confirmed the ellipsoidal shape of the chromosome. These models also showed that the overall structure of the chromosome itself is a helix, with the two arms intertwining to a total of ~1.5 turns. Going by the knowledge that a DNA sequence called the *parS* site, adjacent to the origin, is held close to the cell pole, the authors hypothesised that moving the *parS* site to a different location on the genome should reorient the chromosome. In line with this expectation, a 5C analysis of the *C. crescentus* chromosome with a repositioned *parS* site showed a contact frequency map in which the points of contact of the two diagonals had changed. The first contact point was now the new *parS* site, and the second was at the diametrically opposite point of the chromosome, and not the terminus. Thus, it became clear that the *parS* site, with its deterministic localisation to the cell pole, is the primary parameter determining chromosome orientation. It also suggested that the terminus is not anchored. Gene expression analysis of *C. crescentus* where the *parS* site had been repositioned, in comparison with another where the *parS* was located fairly close to its original locus, showed that changing the orientation of the chromosome did not affect gene expression. The reader is directed to the original paper for a study of their more detailed results, including the high compaction of the chromosome around the *parS* site, and a discussion of the constraints determining specific contacts between the two chromosomal arms.

---

76 The sequencing was performed using the Polony sequencing technology, which is open source and ultra-low cost option for next-generation sequencing. A Polony sequencer can be readily built in a laboratory using standard commercial equipment (or with any sequencer purchased off the shelf), and all software for running the system are available in the public domain. This technology was developed by George Church's laboratory. For more information, refer to http://arep.med.harvard.edu/Polonator/

In a more recent study, Le and co-workers[77] used a genome-scale procedure called Hi-C[78] to derive contact frequencies from 3C libraries of *C. crescentus* swarmer cell chromosomes. In preparing 3C libraries for the Hi-C experiment, ligation of cross-linked DNA after restriction digestion is performed such that these regions are tagged with biotin. Streptavidin beads can then be used to purify only the informative DNA fragments, which link two spatially proximal loci together. Subsequently, paired-end sequencing is performed. Reads are mapped back to the reference genome sequence, and the linking of two chromosomal loci by the same read pair represents evidence of their spatial proximity in the cell. In a manner similar to the 5C experiment, contact frequency maps can be built from Hi-C data as well. The Hi-C-based contact frequency map of *C. crescentus* resembled that from the 5C experiment in showing the cross-diagonal pattern. In addition however, the Hi-C study reports the presence of local contacts, which produce domains such that loci within the same domain are more likely to interact with each other than with those outside the domain. In a contact frequency map, these domains appear as squares along the main diagonal that represents interactions within the same chromosomal arm. A locus at the 5′-end of a domain is more likely to interact with another that is immediately 3′ to it than with one which is equidistant on its 5′ side. Thus contacts across domain boundaries are rare. The *C. crescentus* swarmer cell chromosome can be divided into ~25 such domains in the 30–420 kb size range.[79] The researchers then investigated parameters which determine these structural features, by inducing perturbations and then performing Hi-C experiments on the perturbed cells. Treatment of the bacteria with rifampicin which inhibits transcription elongation, immediately destroyed the domain structure, while retaining interactions across chromosomal arms. Further experiments, in which highly expressed genes were moved to locations with lowly expressed genes, clearly showed that the presence of a region of high gene expression creates a sharp domain boundary in its vicinity. Next, antibiotic-mediated inhibition of DNA supercoiling reduced contact frequencies in the 20–200 kb size range. Deletion of a chromatin-associated protein called HU (described in greater detail in later sections) resulted in a reduction in short-range contacts within the 100 kb range. Finally, loss of a protein called SMC,

77 Le T. B., Imakaev M. V., Mirny L. A. and Laub M. T. 2013. 'High-resolution mapping of the spatial organization of a bacterial chromosome.' *Science*. 342: 731–34.

78 Lieberman-Aiden E., van Berkum N. L., Williams L., Imakaev M., Ragoczy T., Telling A., Amit I., Lajoie B. R., Sabo P. J., Dorschner M. O., Sandstrom R., Bernstein B., Bender M. A., Groudine M., Gnirke A., Stamatoyannopoulos J., Mirny L. A., Lander E. S. and Dekker J. 2009. 'Comprehensive mapping of long-range interactions reveals folding principles of the human genome.' *Science* 326: 289–93.

79 Elegant genetic experiments by the Boccard laboratory have described what are known as macrodomains in the *E. coli* chromosome. See Valens M., Penaud S., Rossignol M., Cornet F. and Boccard F. 2004. 'Macrodomain organization of the *Escherichia coli* chromosome.' *EMBO Journal* 27: 4330–4341.

which enables chromosome compaction, decreased the frequency of inter-arm interactions, and also caused the two arms of the chromosome to lose their colinearity. Therefore, gene expression and supercoiling could be clearly shown to be important determinants of domain boundary formation, whereas the chromosome compacting protein, SMC maintains long-range inter-arm contacts. The researchers also performed Hi-C experiments of swarmer cells during different stages of the cell cycle. At certain points in the cell cycle, while the two daughter chromosomes are segregating, the origin of one chromosome is transiently positioned close to the terminus of the other. This they could reproduce in their Hi-C experiments. However, they also note that the frequencies of these contacts are still much lower than close-range interactions within the same chromosome. This suggested the presence of mechanisms that ensures separation of the two chromosomes such that they do not get entangled during cell division.

5.7.4 RNA polymerase occupancy and the σ-factors

Much of the regulation of gene expression acts by affecting the function of RNA polymerase, the enzyme that catalyses transcription. We will discuss genomic studies of how transcription factors regulate transcription initiation in the next section. Here we look at studies investigating how the occupancy of the RNA polymerase differs from the promoter and the corresponding gene body. We will also consider studies interrogating the regulatory targets of σ-factors and the potential for cross-talk among several σ-factors in the same organism.

5.7.4.1 RNA polymerase stalling and σ-factor retention

Let us assume that the RPO holo enzyme has been recruited to the promoter and is ready to start transcription, after much deliberations among various regulators. Does recruitment to a promoter predict immediate release of RPO from the promoter and subsequent transcription elongation? Or is promoter release and transition to elongation a regulated process in itself? Is the transition from initiation to elongation always accompanied by the release of the σ-factor from the RPO? To answer these questions, Reppas and colleagues from the laboratories of George Church and Kevin Struhl performed a genome-wide ChIP-chip analysis of RPO and the major σ-factor σD in *E. coli.*[80] For their experiments, the authors used a high density genome tiling microarray containing over 380,000 50-mer probes. First, the authors performed a ChIP-chip analysis of RPO, using an antibody against a subunit forming part of the core enzyme. This along with a transcriptome of *E. coli* under the same conditions, and constructed using the

[80] Reppas N. B., Wade J. T., Church G. M. and Struhl K. 2006. 'The transition between transcription initiation and elongation in *E. coli* is highly variable and often rate limiting.' *Molecular Cell* 24: 747–57.

same microarray platform, served as an early experimental genome annotation study for this model organism. Their experimental annotation resulted in >1,800 transcribed fragments or 'transfrags'. A ChIP-chip experiment against σD was also performed. Binding regions or peaks identified from the σD ChIP-chip data could be expected to represent promoter sites. A total of ~1,300 promoters were identified.

For all transcribed regions, as identified from their transcriptome experiment, the researchers compared the ChIP-chip signal for the RPO across the promoter and the transcript. In general, they found that the RPO occupancy was higher in the promoter than across the body of the transcript. Importantly, Reppas and co-workers noted RPO and σD occupancy at promoters with no detectable transcriptional activity under the condition tested. This observation could be validated using ChIP-qPCR for selected genes and promoters. Though the ~300 σD peaks associated with non-transcribed regions had lower ChIP signal than those adjacent to transcribed regions, there was a large overlap between the signal distributions for the two classes of peaks; this suggested that the ChIP signal for a σD peak is not sufficiently predictive of transcriptional activity. Thus, the authors could conclude that RPO is often poised at promoters, even if not actively transcribing. It is believed that poised RPO at promoters can quickly initiate transcription at these promoters when the conditions become permissive. However, it is notable that Jay Hinton and colleagues[81]–using ChIP-chip of RPO in stationary phase *Salmonella enterica Typhimurium*–observe no poised RPO upstream of genes that are rapidly activated following transfer to fresh growth media; this group therefore suggests that rapid, early transcriptional adaptation during the lag phase of growth involves '*de-novo*' repositioning of RPO, and not activation of poised RPO. Similarly, a ChIP-chip study in *Bacillus subtilis* showed an even RPO occupancy from the promoter to the corresponding coding sequence;[82] where promoter proximal accumulation of RPO turned out to be candidates for transcriptional control by attenuation.[83]

Let us return to the study by Reppas and colleagues. The next question was whether promoters where RPO was poised and not immediately transitioning to elongation were just difficult to unwind. For this, the researchers calculated the melting temperatures (T_m) of the DNA duplex around promoters and compared

---

[81] Rolfe M. D., Rice C. J., Lucchini S., Pin C., Thompson A., Cameron A. D., Alston M., Stringer M. F., Betts R. P., Baranyi J., Peck M. W. and Hinton J. C. 2012. 'Lag phase is a distinct growth phase that prepares bacteria for exponential growth and involves transient metal accumulation.' *Journal of Bacteriology* 194: 686–701.

[82] Ishikawa S., Oshima T., Kurokawa K., Kusuya Y. and Ogasawara N. 2010. 'RNA polymerase trafficking in Bacillus subtilis cells.' *Journal of Bacteriology* 192: 5778–5787.

[83] Regulation of transcription by premature termination, arise from simultaneous transcription and transcription; example: The classical textbook case of the trp operon in *E. coli*.

this property across promoters with poised and rapidly-elongating RPO. They did observe a higher T_m trough[84] for promoters with poised RPO than for those associated with expressed transcripts. Thus, ease of DNA unwinding is a parameter influencing RPO stalling at promoters.

Finally, ChIP-chip of σD showed that σD is rarely associated with the elongating polymerase. This was based on the observation that binding signal could be observed only at promoters. Nevertheless, in a few cases, σD binding signal was noticeable well inside the body of the transcript. Earlier studies from the Struhl laboratory had shown that this could occur in a condition-dependent manner.[85] Such trafficking of σD with the elongating RPO may permit rapid rebinding of the holo enzyme to the promoter, thus re-initiating the transcription cycle. That said, such events are likely to be rare.

5.7.4.2 A multitude of σ-factors and their targets

Many bacteria encode multiple σ-factors. *E. coli* has seven σ-factors (Fig. 5.4), whereas the soil bacterium *Streptomyces coelicolor* has over a 60.[86] These belong to three general families: The σ70 family includes the essential, major σ-factor, σD and several types of alternative σ-factors; the σ54 family is the second class of σ-factors, where open complex formation typically requires active unwinding aided by ATPases; the extracytoplasmic σ-factor (ECF) family is divergent in sequence from the above classes of σ-factors and its members are regulated through sequestration by anti-σ-factors. Most bacteria encode a recognisable homolog of σD factor, and optionally other alternative σ-factors from the same σ70 family. Most bacterial genomes encode either one or no member of the σ54 family–rarely seen are genomes encoding two σ54 members. In many organisms including the above-mentioned champion *S. coelicolor*, ECF σ-factors easily outnumber the σ70 and σ54 members. The number of ECF σ-factors varies enormously among genomes, and is likely a reflection of an organism's lifestyle.

The partitioning of the gene regulatory space among multiple σ-factors still remains poorly understood. In organisms with multiple σ-factors, what is the overlap between the targets of these regulators? Do different σ-factors regulate distinct sets of genes? Or do different σ-factors regulate overlapping sets of genes, reflecting the requirement of a given target under multiple stress conditions? To what extent does the observation of a binding event of a σ-factor to a promoter

---

84 A local T_m minima within a promoter.

85 Wade and Struhl. 2004. 'Association of RNA polymerase with transcribed regions in *Escherichia coli*.' *Proceedings of the National Academy of Sciences USA*. 101: 17777–17782.

86 See the following short paper for a brief survey of σ-factors in bacterial genomes. Kill K., Binnewies T. T., Sicheritz-Pontén T., Willenbrock H., Hallin P. F., Wassenaar T. M. and Ussery D. W. 2005. 'Genome update: Sigma factors in 240 bacterial genomes.' *Microbiology*. 151: 3147–3150.

reflect transcription of that gene? Below, we will discuss two studies investigating these issues.

Wade and colleagues performed a ChIP-chip analysis of the heat-shock σ-factor, σH (belonging to the σ70 family) in *E. coli*, using a relatively low-resolution genome tiling array[87] in which a 60-mer probe was placed with an average spacing of 223 bp. They identified ~90 (with an upper limit of ~130) peaks of σH binding, which should represent promoters from which a σH-containing RPO holoenzyme initiates transcription. These binding sites agreed well with a previously-published transcriptome study of σH-dependent gene expression. The first main finding was that ~25% of σH binding sites were within ORFs or in inter-genic regions between converging genes where transcription is not known to initiate. That intra-genic binding could be verified by ChIP-qPCR experiments showed that these findings were not an artifact of the microarray experiment or their data analysis procedures. Bioinformatic analysis predicted the presence of σH promoter signals within these ORFs. Association at these regions was also dependent on heat shock, showing that these regions did not simply show up in any ChIP experiment. Further ChIP experiments showed that RPO was bound to regions downstream of the predicted intragenic promoter, but not upstream, showing that these sites were likely to be bonafide sites for transcription initiation. Based on the absence of a Shine Dalgarno site downstream of the promoter, the researchers propose that these transcripts could represent novel non-coding RNA whose expression is dependent on heat shock and σH. We note here that some of these findings had been flagged as artifacts of the ChIP methodology by Waldminghaus and Skarstad,[88] whose concerns were recently countered by Wade and co-workers.[89] Next, the researchers compared their list of σH targets with those of σD identified in their previous studies. They found that a majority of σH targets were also bound by σD. *In-vitro* transcription assays showed that these genes could indeed be transcribed equally well by both σH and σD, and from the same start sites. The researchers also note that many promoters bound by σH have sequences with striking similarities to the bipartite structure of the classical σD promoters. Similarly, a comparison, by these researchers, of previously identified targets of another alternative σ-factor, σE showed a strong overlap between its targets and those of σD. Many targets of alternative σ-factors have niche-specific roles and are transcribed only by the appropriate alternative σ-factor. However, several others could be required under multiple conditions, and their ability to be transcribed by the major as well as an alternative σ-factor could allow the latter to 'augment' transcription from the former.

---

87 Compared with the 12 bp resolution microarray used by Reppas and co-workers, whose work on poised RPO is described in the section above.

88 Waldminghaus and Skarstad. 2010. 'ChIP on chip: Surprising results are often artifacts.' *BMC Genomics* 11: 414.

89 Bonocora R. P., Fitzgerald D. M., Stringer A. M. and Wade J. T. 2013. 'Non-canonical protein-DNA interactions identified by ChIP are not artifacts.' *BMC Genomics* 14: 254.

Geobacter sulfurreducens is a gram-negative bacterium, which can be used to produce electricity from organic matter. It has a 3.8 Mb genome, predicted to encode ~3,700 genes (~3,400 protein-coding genes). Qiu and colleagues had performed an experimental annotation of the genome of *G. sulfurreducens,*[90] along the lines of the study of *E. coli* by Cho and colleagues described earlier in this chapter. This study identified ~1,400 operons/transcription units in the *G. sulfurreducens* genome. In a more recent study, Qiu and co-workers used ChIP-chip to describe the binding sites of four different σ-factors encoded by the *G. sulfurreducens* genome.[91] This organism contains six σ-factors. The researchers left two of these out as they could not identify the right conditions for their induction or for the expression of their known targets. The four σ-factors assayed in their study included the major σ-factor σD, the heat stress σH, the nitrogen-limitation associated σN and the starvation/stationary-phase related σS. These experiments were performed under multiple conditions, including those associated with the activity of specific σ-factors. In total, the researchers identified over 1,500 σ-factor-binding regions. As many as 60% of all genes were bound by the major σ-factor, σD. Each of the three alternative σ-factors interrogated here are bound to promoters for ~25% of genes. These show that σD is responsible for much of the transcription, as expected from the large body of previous σ-factor literature. However, the number of targets for each alternative σ-factor was surprisingly large.[92] The target gene list for one σ-factor overlapped considerably with that of another, though there were several targets unique to each σ-factor (Fig. 5.19). The study also showed that each of the three alternative σ-factors regulated the major σD, with each σ-factor being auto-regulatory as well. Using what is called flux balance analysis–a computational method to analyse metabolic networks and predict growth rates under various cellular conditions–the authors predicted genes which were essential for growth. These genes could be those that are essential, irrespective of the environmental condition; and those that are essential in a condition-dependent manner. Nearly every universally-essential gene was regulated by σD, either solely or with at least one other σ-factor; however several condition-specific essential genes were regulated by the alternative σ-factors. Though different environmental conditions resulted in large changes in gene expression, the binding profiles of σ-factors remained almost constant across these

---

90 Qiu Y., Cho B. K., Park Y. S., Lovley D., Palsson B. O. and Zengler K. 2010. 'Structural and operational complexity of the *Geobacter sulfurreducens* genome.' *Genome Research* 20: 1304–1311.

91 Qiu Y., Nagarajan H., Embree M., Shieu W., Abate E., Juárez K., Cho B. K., Elkins J. G., Nevin K. P., Barrett C. L., Lovley D. R., Palsson B. O. and Zengler K. 2013. 'Characterizing the interplay between multiple levels of organization within bacterial sigma factor regulatory networks.' *Nature Communications* 4: 1755.

92 The number of binding sites for σH was ~275, compared to 80–130 for σH in *E. coli*, as described by Wade and colleagues.

conditions. These suggested that changes in gene expression were not explained by repositioning of σ-factors, but possibly by differential activation of appropriate promoters by other players (such as transcription factors). An interesting manner in which the σ-factor network of *G. sulfurreducens* differed from that of *E. coli* was the high expression of σS (which is expressed during starvation in *E. coli* ; see section below) during rapid growth. As can be expected to follow from its high expression, σS was bound to as many genes during exponential growth as it was during starvation. The paradigm of σN is in the regulation of the nitrogen-starvation response. However, in *G. sulfurreducens*, σN seemed to have an expanded role, including the regulation of many genes involved in central energy metabolism.

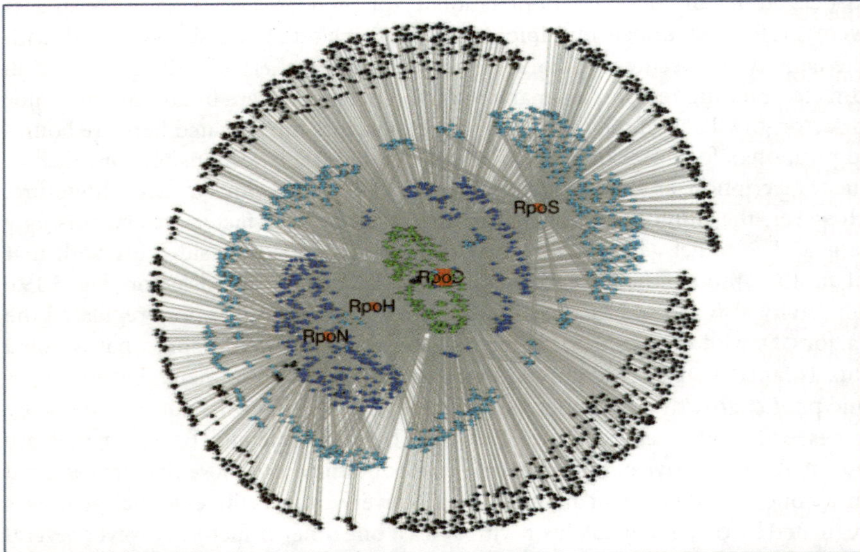

Figure 5.19 *The σ-factor regulatory network of Geobacter sulfurreducens.* In this figure, binding interactions between σ-factors and their target genes are marked by grey lines. The four σ-factors are represented by red squares, with the size of the square indicative of the number of genes regulated by the σ-factor. Note the large number of targets of each of the alternative σ-factors, based on the number of grey lines emerging from them. Genes in green, located close to the centre of the network, are bound by all four σ-factors; those in dark blue by three; those in light blue by two; and those in black by a single σ-factor. This was drawn using the Cytoscape network-rendering software.

The above studies had investigated interactions among distinct types of σ-factors. However, some members of α-proteobacteria encode multiple homologs of the σH group of σ-factors, which generally regulate the heat-shock response. An example is the bacterium *Rhodobacter sphaeroides*, which encodes two heat-shock σ-factors, σH1 and σH2. Dufour and colleagues from Tim Donohue's laboratory used ChIP-chip and gene expression microarrays to study the features of gene regulation by σH1 and σH2 in *R. sphaeroides*.[93] The ChIP-chip analysis[94] of the two heat-shock σ-factors–using a custom-made genome tiling microarray–resulted in a large number of binding sites: ~1,100 for σH1 and ~1,700 for σH2. These very large numbers[95]–for a genome of ~4.6 Mb[96]–are not discussed much. However, microarray analysis of genome-wide gene expression showed that only ~240 and ~180 genes were down-regulated in σH1$^-$ and σH2$^-$ strains, when compared to σH1$^+$ and σH2$^+$[97] respectively. The overlap between the genes differentially expressed in σH1$^-$ or σH2$^-$ and those bound by the respective σ-factor was then considered to represent the targets of the two σ-factors. Finally, ~45 genes were defined as regulated by both the σ-factors, ~130 only by σH1 and ~100 only by σH2. Many of the genes regulated by both the σ-factors were involved in core cellular processes including electron transport, cell membrane integrity and DNA repair. Genes regulated only by σH1 appeared to represent a classical heat-shock response. However, genes regulated only by σH2 included many proteins involved in oxidation–reduction reactions. In line with this observation, σH2 expression was induced after the introduction of oxidative stress. Sequence alignments and motif searches for DNA sites bound by the two σ-factors showed that the two proteins bound similar sequences, but with very clear differences at individual base positions. Thus, it is possible that a gene duplication event producing two σH-factors, followed by divergence, could have established a certain level of 'convergence' between the transcriptional responses of *R. sphaeroides* to heat shock and to oxidative stress.

---

93 Dufour Y. S., Imam S., Koo B. M., Green H. A. and Donohue T. J. 2012. 'Convergence of the transcriptional responses to heat shock and singlet oxygen stresses.' *PLoS Genetics* 8: e1002929.

94 In general, σ-factor ChIP experiments are performed using antibodies raised directly against the protein; this was done for most of the *E. coli* σ-factor studies, as well as for *G. sulfurreducens*. However, for *R. sphaeroides,* the researchers were able to produce an antibody against only one of the two σ-factors. The other σ-factor was tagged on a plasmid with an epitope and an antibody against the epitope used to perform the immunoprecipitation reaction. However, the authors do show that the epitope tag does not affect the function of the protein by comparing the gene expression profiles of a strain expressing the endogenous σ-factor and another expressing the tagged σ-factor on plasmids.

95 Compared to the numbers discussed above for *E. coli* and *G. sulfurreducens*.

96 Spread across two chromosomes and five plasmids.

97 In the σH1$^+$ and σH2$^+$ strains, the σH-factors were expressed ectopically in the respective σH-backgrounds, accessed on 10th July, 2014.

5.7.4.3 The σS of general stress response in *E. coli*

Among the several σ-factors in *E. coli*, one that has attracted considerable interest is σS, which is responsible for the transcription underlying the general stress response. During batch growth, the expression of σS is low during the exponential phase, but increases considerably on entry into the stationary phase. It is also induced by other stresses such as osmotic and acid stress. In order to succesfully transcribe genes under its control, σS must first compete effectively with σD for binding to the core RPO; this is a considerable challenge as the affinity of σS to the core RPO is much less than that of σD to the core, and the level of σD overwhelms that of σS even under starvation or other stress conditions. σS comes under complex regulation at the transcriptional, post-transcriptional and the post-translational level; and the competition between σS and σD is defined not only by their expression levels, but also by a multitude of other factors that post-translationally affect the availability of the two σ-factors or the holo enzyme containing these σ-factors.

In order to systematically catalogue the genes and gene functions regulated by σS, and the promoter sequences bound by the σS-containing RPO, Weber and colleagues from Regine Hengge's laboratory performed a microarray-based gene expression analysis of a strain of *E. coli* carrying a transposon insertion (and therefore inactivation) in σS, and compared it with a wild type strain with an intact σS gene.[98] Genes which are differentially expressed in the σS$^-$ strain, when compared to the σS$^+$ control, were defined as being regulated by σS. σS being an activator of transcription, many genes should be down-regulated in the σS$^-$ strain when compared to the wild type control. However, it is important to note that this approach will produce a list of genes, which are indirectly regulated by σS as well. For example, σS might regulate a gene *x*, which in turn may control the expression of gene *y*. Though gene *y* may not be directly under σS control, it will be flagged as differentially expressed and therefore regulated by σS. One could expect, for example, genes that are up-regulated in σS$^-$ relative to the wild type to be somehow under indirect control of σS, perhaps being repressed by a gene directly regulated by σS. Distinction between direct and indirect effects can be made using a ChIP-chip/-seq study of σS, but this was not done in the study by Weber and co-workers. To produce a comprehensive list of σS targets, these researchers performed their experiments under three different conditions (stationary phase in rich medium, osmotic stress and acid stress), all known to induce σS expression. Across the three conditions sampled, a total of ~480 genes were down-regulated in the σS$^-$ mutant by at least two-fold over the wild type control. These were defined by the researchers as σS-controlled genes (directly

[98] Weber H., Polen T., Heuveling J., Wendisch V. F. and Hengge R. 2005. 'Genome-wide analysis of the general stress response network in *Escherichia coli*: σS-dependent genes, promoters and sigma factor selectivity.' *Journal of Bacteriology* 187: 1591–1603, accessed on 10th July, 2014.

or indirectly). Consistent with the fact that σS is an activator of gene expression, only ~90 genes were up-regulated in the mutant. As mentioned above, these genes might be repressed by a direct target of σS, or be transcribed by one of the other σ-factors as a consequence of relief from competition with σS. Among the 480 σS-controlled genes, only 140 were consistently down-regulated in $σS^-$ across all the three conditions. These 140 genes were called the core σS targets. Thus a considerable proportion of the transcriptional effects of σS inactivation is condition-specific; in particular, most of the condition-specific effect was accounted for by one treatment–growth under osmotic stress. Though σS-controlled genes were spread across the chromosome, there were a few clusters of σS targets including one with multiple acid stress–responsive genes, as well as a large region (~10% of the genome) around the origin of replication with virtually no σS targets. Alignment of sequences upstream of the 140 core σS targets revealed the presence of a TCTATACTTAA motif, which agreed well with an extended –10 sequence previously discussed as a promoter for σS. On the other hand, condition-specific σS targets did not show a significant presence of this motif. This suggested to the researchers that transcriptional activation by σS at condition-dependent promoters might occur at suboptimal sites, and possibly require accessory regulators. Detailed examination of gene functions regulated by σS showed that several genes for anaerobic respiration were down-regulated in $σS^-$, whereas those for aerobic respiration were up-regulated. This suggested that σS might have a role in making a switch from aerobic to fermentative energy metabolism during starvation.

As mentioned earlier in this section, the mechanisms that enable σS to compete with the dominant σD is a matter of debate. Regulators–such as the non-coding 6S RNA and the protein Rsd–that increase the competitive edge of σS go up in expression during the transition to stationary phase, and typically sequester σD or the RPO holoenzyme bound to σD. This in effect is believed to reduce the availability of σD for transcription, thus giving more breathing space for σS to act. A microarray study of the effects of a 6S RNA inactivation on gene expression found patterns that were consistent with the biochemical function of this RNA: Specific effects on gene expression from promoters with a weak –35 element, though not necessarily the targets of σS.[99] However, Rsd inactivation did not recover any differentially expressed genes. We will leave these interesting cases alone here and focus instead on a second mode of regulating σ-factor competition. This is represented by a protein called Crl, which unlike the above-mentioned regulators of σ-factor competition, binds to σS. A study by Typas and co-workers, again from the Hengge laboratory, showed that the Crl protein increased the ratio of the transcriptionally competent RPO-bound-σS to free σS,

[99] Cavanagh A. T., Klocko A. D., Liu X. and Wassarman K. M. 2008. 'Promoter specificity for 6S RNA regulation of transcription is determined by core promoter sequences and competition for region 4.2 of σ70.' *Molecular Microbiology* 67: 1242–1256, accessed on 10th July, 2014.

while decreasing the corresponding ratio for σD.[100] Thus, Crl is a direct, positive regulator of σS activity. The researchers performed a microarray-based gene expression analysis of a Crl⁻ strain and compared it with the wild type. About 30–80 genes–depending on the fold-change threshold–were differentially expressed in the Crl⁻mutant. Almost all these genes were down-regulated in the mutant. Unlike studies on molecules targeting σD or RPO-σD, ~90% of these genes were known targets of σS. Further microarray investigations showed that the effects of Crl on gene expression disappeared in a σS⁻ background. These showed that all the transcriptional effects of Crl were mediated through σS, and that Crl is a major player in determining the competition between σS and σD.

5.7.5 Transcription factors and transcriptional regulatory networks

Transcription factors (TF) are probably the most extensively studied family of regulators in bacteria. They are a diverse group of proteins, most of which contain a helix–turn–helix (HTH) DNA-binding motif. As mentioned in Chapter 2, the number of TFs encoded in a genome is proportional to the square of the genome size (or the total number of genes in the genome). Some of the reduced genomes have a few TFs, with the genome of *M. pneumoniae*[101] predicted to code for only eight TFs. The large genome of *Streptomyces coelicolor* could code for nearly a 1,000 TFs, which is only ~25% less than the number predicted for the human genome. One can safely say that most genomes of free-living bacteria encode a complex repertoire of TFs, which enable them to respond to a variety of environmental and cellular stimuli. This section will focus almost entirely on TFs encoded in the genome of *E. coli*, and where relevant, in comparison with those of the related pathogen, *Salmonella enterica*. We will discuss two types of studies. The first will represent quasi-genomic computational and statistical analysis of regulatory networks available in databases such as RegulonDB. This has been an active field for a while and has mushroomed a large number of papers. We discuss only a very small selection of those studies which are illustrative of our ability to synthesise genome-scale data–from diverse sources–into meaningful findings; whether these findings reinvented the wheel or not is for us to discuss another day. The second set of papers will either use computational approaches on publicly available large-scale data towards constructing networks of TF–target gene interactions, or generation of genomic information that help study specific regulators or regulatory interactions. Before we get into these studies which interrogate the targets of TFs, we will present a survey of *E. coli* TFs.

---

[100] Typas A., Barembruch C., Possling A. and Hengge R. 2007. 'Stationary phase reorganisation of the *Escherichia coli* transcription machinery by Crl protein, a fine-tuner of σS activity and levels.' *EMBO Journal* 26: 1569–1578, accessed on 10th July, 2014.

[101] Which was experimentally annotated by Guell and colleagues, and shown to display complex patterns of gene expression, not obviously explained by its few TFs, accessed on 10th July, 2014.

5.7.5.1 The nature of transcription factors in *Escherichia coli*, as determined from their sequences

TFs encoded by a genome can be predicted using sequence comparison methods. One normally starts off by compiling a list of sequence families that represent TFs. Multiple sequence alignments of representative sequences from such families are available in public databases such as PFAM[102] or CDD (Conserved Domains Database).[103] Whole proteome sequence datasets can be scanned against such alignments using programs from the HMMER suite[104] or RPS-BLAST.[105] Several attempts have been made to predict TFs (and other regulatory proteins) in bacterial genomes. Some of these have been transformed into publicly available databases–these include DBD[106] by Sarah Teichmann's group, MiST (Microbial Signal Transduction)[107] by Igor Zhulin and colleagues, or in unnamed collections such as that built by Michael Galperin.[108] A range of studies have investigated the conservation of particular types of TFs across bacteria and across kingdoms of life.[109] We do not discuss these here; instead we focus on the characteristics of TFs encoded by the *E. coli* genome.

The genome of *E. coli* (K12 MG1655) is predicted to encode ~300 TFs, the exact number depending on false positive–true negative tradeoffs in the prediction algorithm. In most cases, the DNA binding segment of the TF is a HTH (Helix-Turn-Helix) motif. However, many TFs are 'two-headed' in the sense that the DNA-binding domain is present alongside a second domain of a different function. As many as 80–85% of all TFs in *E. coli* are predicted to be two-headed. Nearly half of all *E. coli* TFs have a small molecule-binding domain alongside the DNA-binding domain. This is exemplified by the classical Lac repressor, in which the binding of allo-lactose to the small molecule-binding domain abrogates binding of the TF to the DNA. A particular family of TFs–called the LysR family–has over 40 members in *E. coli* and has an HTH domain

---

[102] http://pfam.sanger.ac.uk, accessed on 10th July, 2014.
[103] http://www.ncbi.nlm.nih.gov/Structure/cdd/cdd.shtml, accessed on 10th July, 2014.
[104] http://hmmer.janelia.org/
[105] http://www.ncbi.nlm.nih.gov/Structure/cdd/wrpsb.cgi
[106] http://www.transcriptionfactor.org
[107] http://mistdb.com/
[108] http://www.ncbi.nlm.nih.gov/Complete_Genomes/SignalCensus.html
[109] See for example: (a) Charoensawan V., Wilson D. and Teichmann S. A. 2010. 'Genomic repertoires of DNA-binding transcription factors across the tree of life.' *Nucleic Acids Research* 38: 7364–7377; (b) Charoensawan V., Wilson D. and Teichmann S. A. 2010. 'Lineage-specific expansion of DNA-binding transcription factor families.' *Trends in Genetics* 26: 388–93; (c) Lozada-Chávez I1., Janga S. C. and Collado-Vides J. 2006. 'Bacterial regulatory networks are extremely flexible in evolution.' *Nucleic Acids Research* 34: 3434–3445.

associated with a second domain predicted to bind to small molecules.[110] Another 10% of TFs contain what is called a response regulator or a receiver domain. This domain gets phosphorylated at an aspartate by a cognate histidine kinase. The activity of these TFs is regulated by phosphorylation. Only about 10–15% of TFs clearly do not have a second domain; even so, some DNA binding domains by themselves include the HTH and a separate region for small molecule binding, as in the repressor of tryptophan biosynthesis operon. A few TFs have an enzymatic domain alongside the DNA-binding domain: An example is the regulator of biotin metabolism genes. The remaining TFs may not have a second domain with detectable homology to known protein families, but have a long-enough stretch of unannotated sequence to have a second unknown domain/function. Thus, the activities of a large proportion of TFs is regulated post-translationally by small molecule binding or phosphorylation.

5.7.5.2 Targets of transcription factors: Mode and scope of action

As mentioned above, RegulonDB has been a long-standing resource for transcriptional regulation in *E. coli*. A class of studies that emerged, with this database as a foundation, is that representing statistical analysis of transcriptional regulatory networks. A transcriptional regulatory network is defined as a set of interactions of a TF with a target gene or operon (Fig. 5.20). Each interaction could be qualified by other information such as the sequence and the location of the binding site, the mechanism of activation or repression, and the environmental and the cellular conditions in which it is switched on. Earlier in this chapter, we had briefly mentioned a few general mechanisms by which TFs activate or repress gene expression.[111] On the basis of those descriptions, it is possible to hypothesise that knowing the binding site location of a TF–relative to the promoter or the TSS–we might be able to reasonably predict the mode of action of the TF on that target gene. In an early bioinformatic survey, Madan Babu and Sarah Teichmann set out to build some descriptive statistics of the locations of TF binding sites, relative to the TSS, and to make associations with the mode of action of the TF.[112] For this, the researchers assembled a dataset of ~1,250 binding sites of ~70 TFs to ~300 promoters representing ~530 genes–the dataset had contributions from RegulonDB and an unnamed resource from the literature.[113] Of the ~70 TFs, ~50% were dual regulators, capable of activation or repression depending on the

[110] In a few well-characterised cases, the small molecule appears to be an amino acid. An example is the repressor of arginine biosynthesis.

[111] Browning and Busby. 2004. 'The regulation of bacterial transcription initiation.' *Nature Reviews Microbiology* 2: 1–9.

[112] Madan Babu and Teichmann. 2003. 'Functional determinants of transcription factors in *Escherichia coli*: Protein families and binding sites.' *Trends in Genetics* 19: 75–79.

[113] Shen-Orr S. S., Milo R., Mangan S. and Alon U. 2002. 'Network motifs in the transcriptional regulation network of *Escherichia coli*.' *Nat Genet* May 31(1): 64–68.

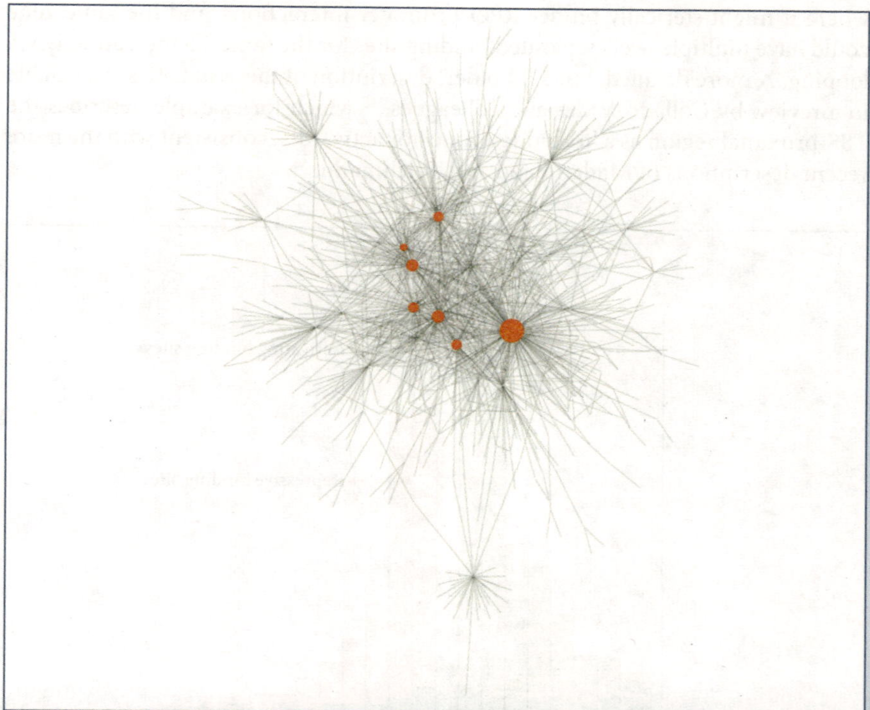

Figure 5.20 *Transcriptional regulatory network of E. coli from the RegulonDB database.* This image of the transcriptional regulatory network of *E. coli* was drawn using the dataset of 1,900 interactions between ~190 transcription factors and ~1,000 target transcription units in the RegulonDB database. The seven global regulators, as defined by Martinez-Antonio and Collado-Vides (2003), are marked by the large, red circles. This was drawn using the Cytoscape network-rendering software.

context, and the remaining 50% almost equally divided between activators and repressors;[114] this information is available in the primary literature and has been compiled together in RegulonDB. By classifying TFs based on their sequence families, the researchers report that knowing the type of DNA-binding domain or the partner domain (second 'head') of a TF does not help to predict whether the TF is an activator or a repressor or a dual regulator. However, the authors describe that most activating binding sites of TFs occur well upstream of the TSS, whereas repressive binding sites could occur both upstream and downstream of the TSS (Fig. 5.21). Patterns of binding site location of repressors is clearly more complex than that of activators: Many binding sites are located downstream of the TSS where it could block RPO elongation; some overlap with the promoter

[114] The current numbers are different: ~40% are dual, 35% are repressors, 25% are activators.

where it might sterically hinder RPO-promoter interactions; and the same gene could have multiple, well-separated binding sites for the same TF indicating DNA looping. A more detailed, but a lot older, description of these statistics is available in a review by Collado-Vides and colleagues,[115] which for example describes the TSS-proximal region as a 'forbidden zone' for activators, consistent with the more recent descriptions of Madan Babu and Teichmann.

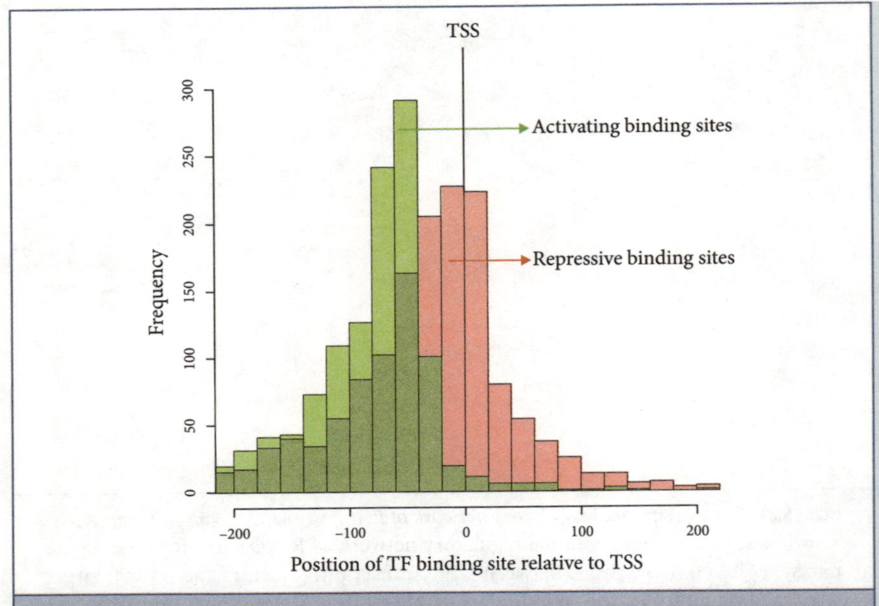

Figure 5.21 *Transcription factor binding site position and type of regulation.* This figure shows the distribution of location of the binding site of a transcription factor to a gene, relative to the transcription start site of the gene, for activating (green) and repressive (red) interactions. Note that repressive interactions tend to be closer to the TSS. DNA looping from distal repressive sites is masked in this figure because of the lower frequency of such sites in the RegulonDB dataset used for this figure. *x*-axis truncated to the −200 to +200 region.

If one ignores the details of TF binding sites on the genome, and focus only on the number and identities of the genes that a TF regulates, we can build a simple transcriptional regulatory network. In such a network, each TF is a source node, and each regulated (or target) gene, a sink node, with an edge between a TF and a target gene representing a transcriptional regulatory interaction, or at

[115] Collado-Vides J., Magasanik B. and Gralla J. D. 1991. 'Control site location and transcriptional regulation in *Escherichia coli*.' *Microbiology and Molecular Biology Reviews* 55: 371–94.

the very least, the binding of the TF to the target gene or its upstream regulatory DNA (Fig. 5.20). As of December 15, 2013, the transcriptional regulatory network of *E. coli* represented in the RegulonDB database includes ~4,200 edges between ~195 TFs and ~1,700 target genes. In addition are ~3,900 edges from the seven σ-factors (individually or in 'OR' combinations) to ~3,000 regulated genes. Most studies tend to separate out regulation by TFs and that by σ-factors, and much of the literature on transcriptional regulatory networks focuses on TF–target gene interactions and less frequently on σ-factor-mediated networks.

A quick analysis of the RegulonDB database shows that ~70% of all TFs regulate ten genes or less; less than 10% have more than 50 targets, with ~5% of TFs regulating more than 100 genes (Fig. 5.22). Thus, most TFs are specific to a few targets–only a few have a broad scope. Both these categories are exemplified by the famous *lac* operon paradigm, where the LacI repressor–that responds to lactose levels–specifically regulates this operon whereas the CRP protein responds to glucose availability and broadly regulates the expression of a variety of catabolic enzymes. Though many target genes appear to be regulated by a single TF (>40%), a majority is bound by multiple TFs,[116] thus suggesting the great extent of combinatorial control. Once again, the *lac* operon–with its two regulator paradigm–exemplifies combinatorial control.

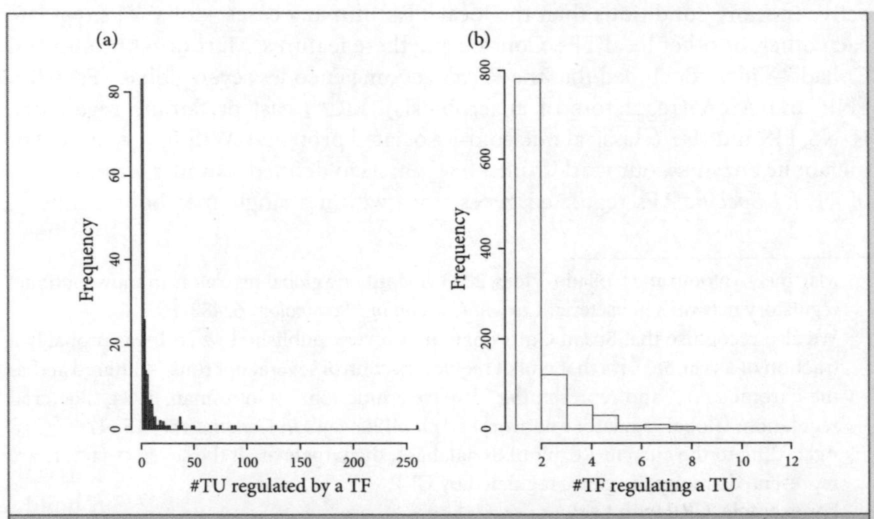

Figure 5.22 *Degree distributions in the transcriptional regulatory network of E. coli.* This figure shows distributions (a) the number of transcription units regulated by a transcription factors; and (b) the number of transcription factors regulating a given gene.

[116] This number could be an underestimation as these data are not complete.

We will first examine the 'out-degree distribution'–representing the number of target genes per TF–in greater detail. From the above description, we can state that there are a few TFs which are global regulators of gene expression, targeting hundreds of genes. Most of the rest are local regulators targeting specific pathways, such as lactose degradation. It is reasonable to derive classifications based on the number of genes a TF regulates–several publications have used such a definition. However, Martinez-Antonio and Collado-Vides established a set of criteria–that go beyond the out-degree–to define global regulators (Figs 5.20 and 5.22).[117] These researchers note that the genes regulated by global TFs should represent multiple functional groups.[118] That the targets of any global TF regulates multiple gene functions is also reflected in the fact that they belong to multiple σ-factor partitions; for example, in the dataset used by these researchers, CRP regulates targets of four different σ-factors.[119] In addition, global TFs regulate their targets in partnership with local TFs,[120] or with other global TFs.[121] This aspect of co-regulation also brings in combinatorial control, which can be investigated using what is known as the 'in-degree' distribution–the number of TFs regulating a given target gene. Aspects of co-regulatory networks, wherein edges are drawn between TFs that jointly regulate their targets, have been discussed in other sources.[122] Further, the researchers also compiled a list of growth conditions in which a TF was known to be active, and found that global regulators tend to be active in more conditions than the local TFs. In many cases, global TFs regulate each other, or other local TFs. Considering these features, Martinez-Antonio and Collado-Vides concluded that the *E. coli* genome encodes seven global TFs–CRP, FNR and ArcA (regulators of anaerobiosis), LRP (feast or famine regulator), H-NS, FIS and IHF (classical nucleoid-associated proteins). Within the context of metabolic enzymes, our work from a few years ago defined a similar classification of TFs.[123] *Specific* TFs regulated genes from within a single metabolic pathway,

---

[117] Martinez-Antonio and Collado-Vides. 2003. 'Identifying global regulators in transcriptional regulatory networks in bacteria.' *Current Opinion in Microbiology* 6: 482–89.

[118] We also recognise that Susan Gottesman, in a review published when I was probably a fraction of a year old says that global regulators control several operons "scattered across the chromosome" and representing "disparate functions" (Gottesman. 1984. 'Bacterial regulation: Global regulatory networks.' *Annual Reviews in Genetics* 18: 415–41.

[119] According to the current RegulonDB database, the targets of all the seven σ-factors are represented among the genes regulated by CRP.

[120] For example, CRP with LacI.

[121] Martinez-Antonio and Collado-Vides use the term "club co-regulation" to describe this.

[122] Balaji S., Babu M. M. and Aravind L. 2007. 'Interplay between network structures, regulatory modes and sensing mechanisms of transcription factors in the transcriptional regulatory network of *E. coli*.' *Journal of Molecular Biology* 372: 1108–1122.

[123] Seshasayee A. S., Fraser G. M., Babu M. M. and Luscombe N. M. 2009. 'Principles of transcriptional regulation and evolution of the metabolic system in *E. coli*.' *Genome Research* 19: 79–91.

whereas 'general' TFs targeted genes from multiple functional categories[124] but with one function being dominant. There are several other ways in which local (or specific) TFs differ from global (or general TFs). For example, the gene expression level of a TF is correlated to the number of binding sites it has–therefore, global TFs are generally expressed at higher levels than local TFs.[125] Targets of global TFs are distributed across the chromosome, whereas those of local TFs are encoded adjacent to the gene encoding the TF itself. Local TFs are encoded close to their target genes probably because of diffusion-limited constraints by which it could be inefficient for a TF with a low expression level to search for a far away binding site.[126] Additionally, it is possible that horizontal transfer of a metabolic pathway alongside its regulator is favoured when the TF is encoded close to its target. Finally, global TFs seem to respond, not directly to signals from the environment, but to signals produced in the cell, presumably in response to an environmental signal.[127] Local TFs might respond to external stimuli or to cellular metabolic products depending on the nature of their target genes.[128]

The above-described definitions of global and local TFs might suggest the existence of a hierarchical structure to the transcriptional regulatory network of *E. coli*. Such hierarchy has been investigated in several studies; however, we discuss one example here. Balazsi and colleagues use the fact that the transcriptional regulatory network of *E. coli* is unidirectional, with an absence of cycles except for autoregulatory rules,[129] to define a hierarchy among TFs (and their targets).

---

124 By definition, broader than a metabolic pathway. For example, arginine biosynthesis as well as tryptophan biosynthesis -which are metabolic pathways -come under amino acid metabolism, which is a functional category encompassing all amino acid biosynthetic and utilisation pathways.

125 Lozada-Chávez I., Angarica V. E., Collado-Vides J. and Contreras-Moreira B. 2008. 'The role of DNA-binding specificity in the evolution of bacterial regulatory networks.' *Journal of Molecular Biology* 379: 627–43.

126 Kolesov G., Wunderlich Z., Laikova O. N., Gelfand M. S. and Mirny L. A. 2007. 'How gene order is influenced by the biophysics of transcriptional regulation.' *Proceedings of the National Academy of Sciences USA* 104: 13948–13953.

127 Martínez-Antonio A., Janga S. C., Salgado H. and Collado-Vides J. 2006. 'Internal-sensing machinery directs the activity of the regulatory network in *Escherichia coli*.' *Trends in Microbiology* 14: 22–27.

128 TFs regulating catabolic enzymes are more likely to sense environmental signals (either from the cell surface or from the cytoplasm after import of the signal), whereas those regulating biosynthetic pathways bind to cellular metabolic products or hybrid signals. Hybrid signals are those that can be importerd into the cell as well as biosynthesised in the cell. See Chapter 3 in my PhD thesis (http://tinyurl.com/aswinthesis).

129 An examination of more recent editions of RegulonDB will show that this is not true any more, and that there are cycles in the network. Every large regulatory network study has used incomplete data (RegulonDB for example), and on many occasions datasets with potentially high error rates (early genomic studies of transcriptional networks in yeasts). Though the authors of such papers (including yours truly) generally try to account for this by artificially removing nodes / edges randomly from the network -or by artificially

The authors identify top-level regulators as those that are not known to be regulated by any other TF in the network.[130] Every other TF is assigned a level in a hierarchical structure, based on the smallest number of interactions separating this TF from a top-level TF. For example, a local TF that is regulated directly by a top-level TF will find itself in the second level of this network, and a third TF regulated only by the above-mentioned local TF will be placed two levels below the top-level TF. Thus, every top-level TF has a set of lower-level TFs and non-TF target genes positioned beneath it. The set of nodes and edges under a top level TF together comprised what Balazsi and colleagues refer to as an *origon*. A total of ~75 origons were defined in this work. Assuming that the top-level TFs are regulated exclusively by their sensing of the environment, it is reasonable to expect that a given environmental stimulus would activate a specific origon, after being sensed by its top-level TF. However, each origon is not isolated, and many origons share several of their lower-tier TFs and targets with other origons. Nearly two-thirds of all origons formed a connected component in the network, sharing members amongst each other. Therefore, a signal activating one of these origons should also trigger lower-tier members of other origons. The remaining one-third seemed to be isolated, and therefore expected to produce highly specific responses to their triggers. By overlaying microarray data on the transcriptional regulatory network, these researchers were able to demonstrate how their classification of the regulatory network shows specific activation of certain origons in response to a given stimulus, and discover that unexpected origons respond to well-researched stimuli.

5.7.5.3 Local organisation of TFs and their targets into network motifs

The division of TFs into global and local, and their organisation into hierarchical modules represent a high-level organisational principle of the *E. coli* transcriptional regulatory network. One can now delve deeper into the network, and search for recurrent patterns of connections involving a small number of nodes (Fig. 5.23). If such patterns exist, what are their properties? Uri Alon's laboratory has pioneered research in this direction, using data from RegulonDB as a starting point.

Using a dataset of ~120 TFs regulating ~400 operons through ~600 regulatory interactions, Shen-Orr and colleagues defined *network motifs* as recurrent patterns

introducing errors in the network and then testing whether their findings hold true for these more incomplete or error-ridden networks. The assumption of randomness in missing data is fraught with difficulties. In most cases missing data are not random errors, but represent biases inherent to the experimental approach or to the interests of the community of researchers; this makes it difficult to assess whether a finding from the analysis of incomplete networks is robust. Therefore, the findings of these studies must be interpreted with care and caution.

[130] The network used by Balazsi and colleagues includes σ-factors as well.

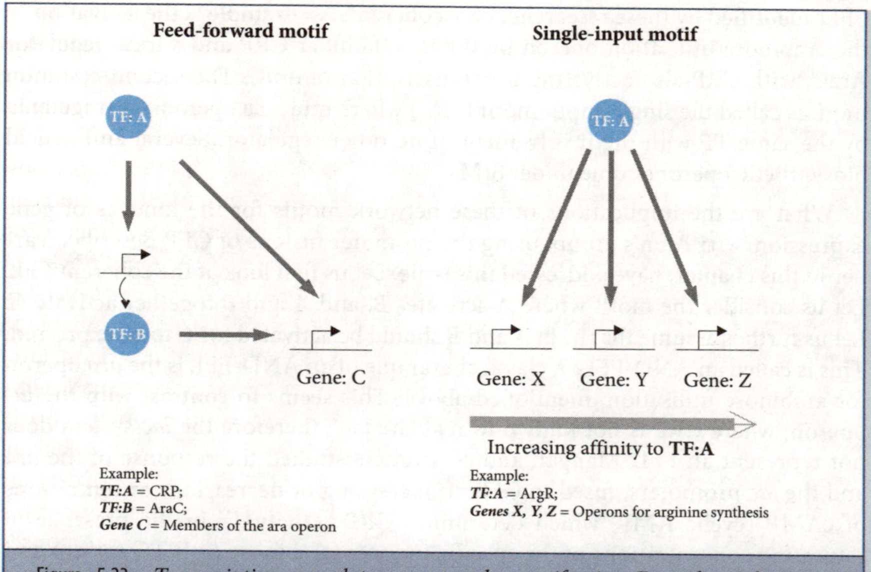

Figure 5.23 *Transcription regulatory network motifs in E. coli.* Schematic representations of the two major types of network motifs (Feed-forward and Single-input) in the transcriptional regulatory network of *E. coli*. The nature of the regulatory interactions (activation/repression) is not indicated here.

of interactions,[131] which occur at frequencies that are higher than expected by random chance, in a regulatory network.[132] A random network was defined by shuffling the interactions among TFs and target genes such that the total number of TFs, that of target genes, and the in-and out-degree distributions of the random network are the same as those of the real network. For all possible patterns involving, say 3 nodes, the frequency of occurrence in the real network can be compared to that in, say 1,000 randomised networks. A pattern can then be flagged as significantly over-represented in the real network if its occurrence in the set of random networks never (<1/1,000 in this case; P-value < 0.001) exceed that in the real network.

The first motif that Shen-Orr and co-workers describe is called the *feed-forward motif* (FFL). In a FFL, TF *A* regulates TF *B* and the two TFs together regulate a third gene *C*. If the effect of *A* on *C* (activation or repression) is the same as the net effect of *A* and *B* on *C*, then the FFL is called coherent; as many as 85% of the

[131] Shen-Orr S. S., Milo R., Mangan S. and Alon U. 2002. 'Network motifs in the transcriptional regulation network of *Escherichia coli*.' *Nature Genetics* 31: 64–68.

[132] The work also discuss clusters of nodes which are densely connected among themselves, with fewer connections beyond. We do not discuss these modules, which the researchers refer to as *Dense Overlapping Regulons (DORs)*.

FFLs identified by these researchers are coherent. An example is the activation of the arabinose utilisation operon by the now-familiar CRP and a local regulator AraC, with CRP also activating the transcription of *araC*. The second common motif is called the single-input motif (SIM) where multiple operons are regulated by the same TF with the involvement of no other regulator. Several amino acid biosynthetic operons come under SIMs.

What are the implications of these network motifs for the kinetics of gene expression? Uri Alon's group, using the promoter fusions of GFP described earlier in this chapter, have addressed this issue. Let us first look at the coherent FFL. Let us consider the motif where *A* activates *B*, and *A* and *B* together activate *C*. Let us further assume that both *A* and *B* should be activated for *C* to be expressed. This is called an AND-FFL. A classical example of an AND-FFL is the *ara* operon for arabinose utilisation, mentioned above. This seems to contrast with the *lac* operon, where CRP is not known to regulate *lacI*; therefore the *lac* system does not represent an FFL. Mangan and co-workers studied the response of the *ara* and the *lac* promoters, fused to GFP, to increasing or decreasing concentrations of cAMP (cyclic AMP, which determines CRP activity),[133] in the presence of saturating concentrations of the specific inducer (arabinose or IPTG for *ara* and *lac* respectively). They plot the promoter activity[134] of the operon of interest against the concentration of cAMP, either when the cAMP level is increased from zero to saturation or decreased from saturation to zero. The researchers observed that the *ara* promoter activity increases more slowly than the *lac* promoter activity with increasing concentrations of cAMP. However, decrease in cAMP levels leads to a rapid and equal rate of decline of promoter activities for both *ara* and *lac* operons. This agrees well with thoeretical calculations which show that the expression of an operon under control of an AND-FFL will be insensitive to rapid fluctuations in the concentration of the inducer of the top-level TF. This is because, a persistent top-level signal is required to activate the longer input arm of the FFL. Since both inputs need to be turned on in an AND-FFL for the downstream gene to respond, the coherent AND-FFL acts as what is called a 'sign-sensitive[135] delay element' in a regulatory network. A contrasting type of coherent FFL is the SUM-FFL where either TF *A* or *B* is sufficient for the activation of *C*, and the magnitude of activation is determined by the sum of the effects of *A* and *B*. An example for this architecture is the flagellar system in *E. coli* where the master regulator FlhDC activates the expression of the σ-factor FliA; the two regulators together activate

[133] Mangan S., Zaslaver A. and Alon U. 2003. 'The coherent feedforward loop serves as a sign-sensitive delay element in transcription networks.' *Journal of Molecular Biology* 334: 197–204.

[134] See earlier Section 5.5 titled '*Gene expression at high temporal resolution using fluorescent reporters*' for a definition.

[135] Sign-sensitive because the response to an increase in signal concentration is different from that to a corresponding decrease in signal availability, accessed on 10th July, 2014.

various operons for flagella biosynthesis. Kalir and co-workers have shown–again using promoter–GFP fusions–that the SUM-FFL,[136] in contrast to the AND-FFL, shows a delay in response to deactivation of the top-level regulator. However, the activation kinetics of the SUM-FFL resembles that of a single-TF variant where FlhDC is the only regulator of flagellar genes.

Finally, let us consider the SIM motif. Many amino acid biosynthetic operons are regulated by a single TF, and thus come under a SIM. Using their promoter–GFP fusion system, Zaslaver and colleagues show that operons under the control of a SIM display just-in-time kinetics.[137] Let us consider two operons *O*1 and *O*2 under the control of a TF *X*. If the regulatory sequences of *O*1 and *O*2 are such that *X* activates *O*1 faster than *O*2, then we achieve a time delay between the activation of the two operons, despite their being under the control of the same regulator. In several amino acid biosynthetic pathways, Zaslaver and colleagues found that genes whose protein products were positioned earlier in a metabolic pathway were activated before those involved in the later steps of the pathway. It also emerged that the maximal promoter activities of the early genes were higher than those of the later genes. The researchers suggest that such precise temporal gene expression patterns ensure that products of the early steps of a metabolic pathway are at high enough concentrations to minimise the effects of dilution by cell growth over the time required for the pathway to produce its final product. Thus, different motifs show distinct kinetic properties, which may have been selected for optimal function in the bacterium's natural habitats.

5.7.5.4 Building transcriptional regulatory networks using genome-scale gene expression data

The most direct way of constructing transcriptional regulatory networks on a genomic scale is by performing ChIP-seq or ChIP-chip experiments for every TF that one is interested in. This can be coupled with knockouts/perturbations of these TFs and followed by the measurement of their impact on gene expression using microarrays or RNA-seq approaches. Both these approaches have been performed for all (or most) TFs in the eukaryote *Sachharomyces cerevisiae*, but to a much more limited extent for *E. coli*.[138] The scale of such an experimental approach is enormous if one were to consider the number of TFs and the number

[136] Kalir S., Mangan S. and Alon U. 2005. 'A coherent feed-forward loop with a SUM input function prolongs flagella expression in *Escherichia coli*.' *Molecular Systems Biology* 1: 2005.0006.

[137] Zaslaver A., Mayo A. E., Rosenberg R., Bashkin P., Sberro H., Tsalyuk M., Surette M. G. and Alon U. 2004. 'Just-in-time transcription program in metabolic pathways.' *Nature Genetics* 36: 486–91.

[138] We have seen examples for both these approaches in the section on σ-factors, and will see more below.

of conditions that one has to sample in order to build a reasonably comprehensive network. However, if one could just generate gene expression profiles for wild type *E. coli* under a carefully selected set of conditions, and then somehow use these data to construct transcriptional regulatory networks, then one brings down the number of experiments required by two to three orders of magnitude. Then, even given the limitations to the confidence one could have in such an indirect, predictive approach, it would be a massive help.

The assumption that one has to make to construct regulatory networks from such gene expression data is that the activity of a TF is dependent on its expression level, and that changes in the expression level of a TF are reflected in those of its target genes. Where such an assumption is valid, the following computational procedure should work (Fig. 5.24). The first step is to assemble gene expression data for the target organism from across a range of environmental conditions. These data will have to be normalised such that they are all on the same scale, and therefore comparable. The result of this step is that each gene is represented by a vector of gene expression measures, with the size of the vector being equal to the number of conditions (times the number of replicates per condition) sampled. Next, measures of similarity between gene expression vectors for all pairs of genes, where one member of each pair is a TF, can be calculated. Finally, pairs of genes for which the correlation is significantly high are identified and edges drawn between them.

Faith and co-workers adopted the above approach for constructing a transcriptional regulatory network for *E. coli* (Fig. 5.24).[139] They assembled gene expression data for *E. coli* generated on a single type of microarray platform, which can be accessed from a publicly-available database called M3D (Many Microbes Microarray Database).[140] All these data were normalised on a common platform. Correlation between gene pairs (one member of each pair being a TF) was calculated using what is called mutual information; the advantage that this measure has over the standard Pearson correlation coefficient is that mutual information does not assume that the two vectors should be linearly related. To measure the statistical significance of the correlation between any gene pair, they calculated the distribution of the mutual information scores for each member of the pair against all other genes in the database; a gene pair was flagged as correlated if the mutual information between that pair was significantly higher than expected from the above-described distribution. This approach, in contrast to one using a fixed threshold, ensures that genes, which for whatever systematic

---

139 Faith J. J., Hayete B., Thaden J. T., Mogno I., Wierzbowski J., Cottarel G., Kasif S., Collins J. J. and Gardner T. S. 2007. 'Large-Scale Mapping and Validation of *Escherichia coli* Transcriptional Regulation from a Compendium of Expression Profiles.' *PLoS Biology* 5: e8.

140 http://m3d.mssm.edu/

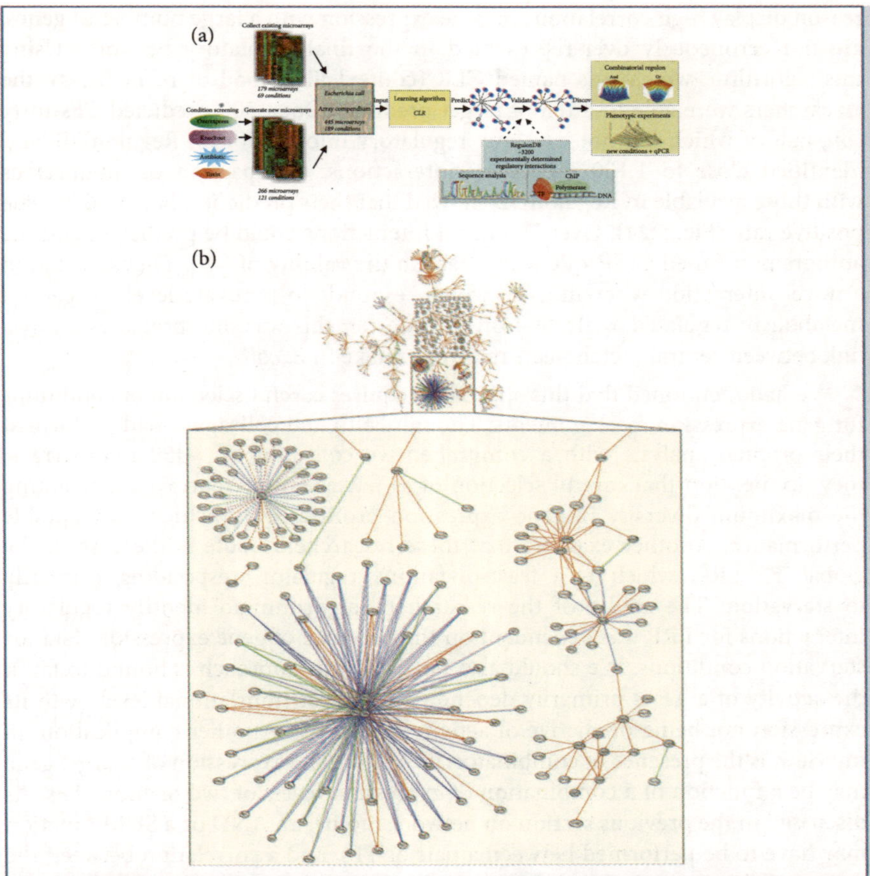

Figure 5.24 *Reconstructing transcriptional regulatory networks from a compendium of gene expression data.* (a) This figure shows a schematic of the procedure by which 'correlations' between gene expression profiles can be used to reconstruct the transcriptional regulatory network of *E. coli*; (b) This figure shows the transcriptional regulatory network of *E. coli* predicted using the above method by Faith and co-workers (2007). Note that this method would perform best for transcription factors such as the regulators of flagella biosynthesis (FlhD and FlhC), whose activities are governed primarily by their expression levels, and not for CRP (a global regulator for which this method predicts only a few targets), whose activity is regulated post-translationally (by binding to cyclic AMP for example), which cannot be captured by these approaches. The two figure panels are reproduced under the Creative Commons Attribution License from Faith J. J., Hayete B., Thaden J. T., Mogno I., Wierzbowski J., et al. 2007. 'Large-scale mapping and validation of *Escherichia coli* transcriptional regulation from a compendium of expression profiles.' *PLoS Biol* 5(1): e8. © Faith et al. 2007.

reason display high correlations in gene expression with a large number of genes, are not erroneously over-represented in the final regulatory network. Using this algorithm, which was named CLR (context likelihood of relatedness), the researchers were able to attempt target prediction for ~300 predicted TFs in *E. coli*, half of which did not have any regulatory information in RegulonDB, and identified close to 1,100 regulatory interactions. Comparison of interactions with those available in RegulonDB showed that their prediction had a ~60% true positive rate (Fig. 5.24). Over 700 novel interactions could be predicted, and the authors performed ChIP-qPCR to establish the validity of ~20. They also found a novel interaction wherein a TF which responds to pyruvate levels in central metabolism regulated a citrate–iron transporter; this was interpreted as a novel link between central metabolism and iron uptake in *E. coli*.

We had mentioned that this approach requires careful selection of conditions for gene expression measurements. Though Faith and colleagues had performed their original analysis with a comprehensive collection of ~450 microarrays, they do mention that careful selection of as few as 60 microarrays representing the maximum diversity in gene expression profiles could achieve comparable performance. Another example that these researchers quote is the case of the global TF LRP, which is a feast-or-famine regulator, responding primarily to starvation. The ability of the researchers' algorithm to identify regulatory interactions for LRP was dependent on the presence of gene expression data for starvation conditions. We should also note that this approach is bound to fail if the activity of a TF is primarily dependent at a post-translational level, with its expression not being predictive of activity (Fig. 5.24). Another complication, in my view, is the presence of combinatorial control. The expression of a target gene may be a function of a combination of expression levels of two or more TFs. As discussed in the previous section on network motifs, an AND or a SUM function may have to be performed between a pair of TFs, and a correlation between the output of this operation and the expression of a target gene computed. Clearly, this would exponentially increase the number of computations, and make the problem challenging. In my experience, one of the most beautiful set of co-regulated genes is the group of operons representing flagella structure, and their regulators FlhDC and the σ-factor FliA. The CLR algorithm recapitulates these interactions exceptionally well. An example in the opposite sense is the global TF CRP, whose activity as a TF is likely to depend on its binding to cAMP and which controls a large number of operons in combination with other regulators. As a possible consequence of both these parameters, very few targets could be recovered for this TF.

5.7.5.5 Building signalling networks with transcriptional outputs: Case studies of two-component phosphorylation networks

At various points in this chapter, we have mentioned two aspects of gene regulatory networks in bacteria. The first is technological: That one can determine regulatory targets of a TF by measuring gene expression changes caused by a perturbation (e.g., knockout) of the TF, albeit with the caveat that we do not differentiate between direct and indirect targets. The second is biological: Most TFs in *E. coli* can be expected to respond post-translationally to a signal, either by binding to it or by being phosphorylated by a sensory histidine kinase. The latter class of TFs, which belong to signalling systems called two-component systems, present an exciting case study for both the above considerations of regulatory networks.

We will first define some typical characteristics of two-component systems in *E. coli*. A two-component system comprises a sensory histidine kinase, which is generally located on the cell membrane, and responds to a signal by autophosphorylating on a histidine residue. This phosphate is then transferred to an aspartate on a second protein called a response regulator or a receiver, which is typically the TF. Most histidine kinases have their own cognate response regulators. In many cases, the gene encoding the histidine kinase and that coding for the response regulator are encoded in the same operon. Though there seem to be many deviations to some of these rules, these are not relevent to our present discussion.

Oshima and co-workers sought to describe the regulatory effects of each of ~30 two-component systems in *E. coli*.[141] Though many two-component systems are expected to perform their primary roles under specific conditions, the scope of the present study covered only standard laboratory rich media. Towards this goal, the researchers generated over 35 deletions in the chromosome. Many strains were deletions of both a histidine kinase and the cognate response regulator, both of which were part of the same operon. In other instances, where the kinase and the receiver were distally located, only one of the two had been deleted. Microarray experiments were performed for these deletion strains, and gene expression in the mutants compared with that in the wild type. The researchers found that over a third of their mutations affected the expression of <20 genes, with a majority resulting in a change in expression of <10 genes. These two-component systems were probably inactive under the standard growth condition used in this study. The remaining two-thirds of the mutants resulted in significant gene expression changes, with four mutants deemed to have global effects on more than 100 genes. These mutants included the previously described global TF, ArcA and its cognate kinase ArcB, which are encoded distally from one another on the chromosome. Many of the targets of these global two-component systems were involved in core cellular processes, which might be reflected in the researchers' observation that these mutants had also displayed severe growth defects. Next, the researchers were interested in discovering the possibility of cross-regulation between two-component systems. By cross-regulation, we mean the possibility of a histidine kinase phosphorylating a non-cognate receiver, or overlap in the target lists of

---

[141] Oshima T., Aiba H., Masuda Y., Kanaya S., Sugiura M., Wanner B. L., Mori H. and Mizuno T. 2002. 'Transcriptome analysis of all two-component regulatory system mutants of *Escherichia coli* K-12.' *Molecular Microbiology* 46: 281–91.

two or more two-component systems. Towards this end, the authors measured the correlation between the expression profiles for every pair of mutants, and prepared a list of mutant pairs which showed high correlation coefficients. Several two component system pairs were found to show correlations that were as high as that found between mutations in ArcA and ArcB, which are members of the same two-component system. We note here that the correlation between mutations in ArcA and ArcB itself was fairly low–relative to what one would expect given the assumption that all the effects of ArcA are caused by its phosphorylation through ArcB and activation of ArcB phosphorylates only through ArcA–suggesting the presence of additional players influencing this pathway. Several genes, which are a target of the stationary phase σ-factor, σ38, and many members of the flagellar system were found to be regulated by multiple two-component systems. The researchers were also able to mention the presence of cascade regulation, where one two-component system regulates another. Once again, we must note that this approach does not distinguish between direct and indirect effects of these regulators, a fact that the researchers admit in their paper. Further, the problem of condition-dependent effects also raises its head here. However, in a different study, Zhou and colleagues performed a *phenotypic microarray* analysis of two-component system mutants, in which respiratory capacities of all these strains could be tested across a panel of several hundred media.[142] This could serve as a screen for identifying conditions in which a TF mutant shows a significant phenotype, which could subsequently be used to perform gene expression studies.

A second angle from which two-component systems benefit our discussion is the dissection of specificity in signal transduction from a histidine kinase to its cognate response regulator(s), as well as the role of bioinformatics in enabling such studies. It had been established previously that much of the specificity, which ensures minimal cross-talk between two-component systems, is at the level of molecular recognition, and therefore contained in the amino acid sequence of the two proteins.[143] Skerker and colleagues from Michael Laub's laboratory set out to identify the sequence features that determine specificity of signal transduction between a histidine kinase and its cognate response regulator,[144] and asked whether such information would help rewire these pathways, such that one histidine kinase can now be engineered to phosphorylate a non-cognate response regulator.

[142] Zhou L., Lei X. H., Bochner B. R. and Wanner B. L. 2003. 'Phenotype microarray analysis of *Escherichia coli* K-12 mutants with deletions of all two-component systems.' *Journal of Bacteriology* 185: 4956–4972.

[143] For example: Skerker J. M., Prasol M. S., Perchuk B. S., Biondi E. G. and Laub M. T. 2005. 'Two-component signal transduction pathways regulating growth and cell cycle progression in a bacterium: A system-level analysis.' *PLoS Biology* 3: e334.

[144] Skerker J. M., Perchuk B. S., Siryaporn A., Lubin E. A., Ashenberg O., Goulian M. and Laub M. T. 2008. 'Rewiring the specificity of two-component signal transduction systems.' *Cell* 133: 1043–1054.

They computationally identified histidine kinases and response regulators encoded in ~200 fully-sequenced bacterial genomes. They took advantage of the fact that many cognate histidine kinase–response regulator pairs are encoded on the same operon. Armed with this knowledge, they assembled a list of ~1,300 interacting histidine kinase–response regulator pairs. The two protein sequences comprising each pair were concatenated into a single sequence, and a multiple sequence alignment performed across these ~1,300 sequences. The objective now was to find pairs of significantly co-varying residues, with one residue in each pair belonging to the histidine kinase and the other to the response regulator. Let us assume that residues *H*1 and *H*2 from a histidine kinase *HK* interact with residues *R*1 and *R*2 in a response regulator *RR*. A second histidine kinase *HK′* has replaced *H*2 with *H*2′, which can interact with residue *R*2′ in *RR′*, but not with *R*2 in *RR* (and vice-versa). Thus, the residue position corresponding to *H*2/*H*2′ covaries with that corresponding to *R*2/*R*2′. A conclusion that can be drawn from this example is that positions corresponding to *H*2/*H*2′ and *R*2/*R*2′ are *specificity determining*. Using mutual information scores to determine columns in the multiple alignment that were co-varying, the authors determined ~40 specificity-determining residues. Though the sequence alignments included the ATP-binding domain of the histidine kinase, which is unlikely to be involved in interactions with the response regulator, ~85% of all specificity determining residue pairs involved the domain that contained the phosphorylated histine in the kinase. Analysis of these residues, in the context of the cocrystal structure of a kinase–response regulator pair, showed that most of these resided on the interaction interface between the two proteins. This gave the researchers the confidence that their predictions were reliable. Next, the researchers replaced the specificity determining residues of one histidine kinase *HK*1 with those of a second kinase *HK*2, and found that *HK*1 now phosphorylated the cognate response regulator for *HK*2 *in-vitro*, leading to the expected rewired transcriptional outcome *in-vivo*. A more recent study, again from the Laub laboratory, has shown that adaptive evolution of specificity determining residues helps minimise crosstalk between two-component systems.[145] Thus, a few residues determine specificity of signal transduction cascades, and modifications at these residue positions can rewire signalling networks.

5.7.5.6 Global transcriptional regulatory networks under small-molecule-binding transcription factors

As mentioned earlier, nearly half of all TFs in *E. coli* contain a small-molecule-binding domain. A straightforward analysis of the domain architecture of *E. coli* TFs, and their known targets in RegulonDB, showed that small-molecule-binding

[145] Capra E. J., Perchuk B. S., Skerker J. M. and Laub M. T. 2012. 'Adaptive mutations that prevent crosstalk enable the expansion of paralogous signaling protein families.' *Cell* 150: 222–32.

TFs are more likely to regulate metabolic enzymes and transporters than non-small-molecule-binding TFs. Therefore, unlike in eukaryotes, where small-molecules signal to TFs through a cascade of signalling pathways, TFs in bacteria tend to bind to small molecules directly and consequently affect a transcriptional response. There are at least two small-molecule-binding TFs in *E. coli*–LRP and CRP–which are global regulators of gene expression. We will use these examples to illustrate the utility of ChIP-chip and ChIP-seq approaches, coupled with gene expression analysis, to understand the genome-wide functions of global TFs. This, we expect, will serve as a foundation for the next section, in which we will discuss various genomic studies of the protein component of the *E. coli* nucleoid and its impact on gene expression.

LRP is a TF that binds to the amino acid leucine, and regulates transcription of a large number of genes. It is a feast-or-famine regulator, which has major roles to play during starvation. For example, the levels of this protein in the cell increases three-fold on transition from exponential growth to stationary phase. Cho and colleagues investigated, using ChIP-chip, the binding profiles of LRP to the *E. coli* chromosome in minimal glucose medium with or without the addition of exogenous leucine.[146] Across three conditions, covering the presence and absence of leucine in exponential growth and stationary phase, the researchers identified ~140 binding regions for LRP. During the mid-exponential phase in the presence of leucine, less than 25% of these sites were occupied; absence of leucine increased the number of binding sites three-fold. The more severe and complex stationary phase environment resulted in LRP occupying an additional 40 sites. Thus, the occupancy of LRP to the genome increased with deprivation of leucine, and more so under severe starvation. Compared to the data available in literature, this ChIP-chip study increased the number of known LRP-bound regions by ~six-fold! Next, the authors examined the effect of LRP on gene expression by two complementary approaches. In the first, they measured changes in RPO occupancy, as measured by ChIP-chip, with the presence or absence of leucine in the medium. They observed a considerable reorganisation of RPO distribution on the chromosome in response to leucine. However, control experiments in which *E. coli* had also been treated with rifampicin showed no such response; recollect that rifampicin is used to identify promoters in a condition-independent manner. The second approach that the authors took was to measure gene expression with microarrays for *E. coli* grown in the presence and absence of leucine. In general, it was found that genes whose expression levels changed in response to leucine levels, also displayed changes in RPO occupancy, asserting the congruence between the two approaches. In addition, Cho and co-workers generated gene expression profiles for a LRP$^-$ mutant strain in the presence and absence of leucine and showed that

[146] Cho B. K., Barrett C. L., Knight E. M., Park Y. S. and Palsson B. O. 2008. 'Genome-scale reconstruction of the Lrp regulatory network in *Escherichia coli*.' *Proceedings of the National Academy of Sciences USA* 105: 19462–19467.

~50% of the transcriptional effects of leucine could be explained by LRP. Finally, by combining the various data that they had generated, the researchers were able to classify genes into various groups depending on the mode of interaction between leucine and LRP: (a) independent, where leucine had no effect on the binding of LRP to the promoter; (b) concerted, where exogenous leucine stimulated the binding of LRP to the promoter; and (c) reciprocal, where leucine relieved LRP–promoter interaction, as observed for most LRP targets. Finally, the targets of LRP extend well beyond amino acid metabolism or even small-molecule metabolism for that matter, asserting the global nature of its function.

A second major global TF that responds to a small molecule is CRP. CRP is activated by binding to cAMP, whose levels increase on glucose deprivation. According to RegulonDB, CRP is the champion TF, regulating the largest number of genes. David Grainger and colleagues from Stephen Busby's laboratory performed a ChIP-chip study of CRP in *E. coli*,[147] in a growth medium lacking glucose; this was one of the first proof-of-principle ChIP-chip studies for bacterial TFs. For this purpose, they used a tiling microarray. As a control–similar to a mock-IP–the researchers performed ChIP-chip using the same antibody in a CRP$^-$ strain. The fold change in intensity between the real IP and that in the CRP$^-$ control was used as measure of the binding signal for CRP to the DNA sequence of the probe. Probes which showed high signals, relative to the overall average signal across the microarray, were defined as corresponding to CRP binding regions. Interestingly, the researchers found that they could describe only <70 binding regions, which is considerably less than what they had expected. Computational analysis had previously identified over 200 high-affinity CRP binding motifs in the *E. coli* genome, and the ChIP-chip study added meagrely to the list of CRP targets already described in RegulonDB. To understand this, the researchers compared the CRP binding signal across the chromosome, with that for CRP in a glucose-containing medium in which the protein should be inactive, and also with that for MelR, a local TF with a single binding site. Interestingly, the researchers noticed that the average background signal for CRP, under conditions in which the protein is active, was considerably higher than that for the inactive protein or for the local TF MelR. This suggested to the researchers that this protein could have a large number of low-affinity binding sites across the chromosome, and that these sites could help shape the chromosome–in agreement with computational predictions that had identified over 10,000 low-affinity CRP binding sites on the genome. Therefore, CRP, unlike a typical TF, could be a nucleoid-associated protein (see below). The researchers also noted that there was a 1 Mb region in the genome that showed low CRP binding signals; this was interpreted as suggestive

[147] Grainger D. C., Hurd D., Harrison M., Holdstock J. and Busby S. J. 2005. 'Studies of the distribution of *Escherichia coli* cAMP-receptor protein and RNA polymerase along the *E. coli* chromosome.' *Proceedings of the National Academy of Sciences USA* 102: 17693–17698.

of the possibility that the chromatin state in this region was non-permissive to CRP binding.

Let us take this opportunity to review the transcriptional response of *E. coli* to a change in carbon source from the preferred glucose to one of lower quality. In the canonical model, such a transition leads to the activation of the genes involved in metabolism of the new carbon source. This transition could also be observed in the RPO ChIP-chip study of Grainger and colleagues,[148] where addition of IPTG, an inducer of the *lac* operon, resulted in an increased RPO occupancy of the *lac* promoter. However, can there be a more global control of gene expression here, especially since carbon source changes result in alterations to growth rate and therefore physiology? Liu and colleagues addressed this issue by performing transcriptome experiments–using microarrays–of *E. coli* grown in glucose and in various other carbon sources supporting progressively less growth.[149] They found that the number of genes that were differentially expressed, in comparison to growth in glucose, increased with decreasing quality of the carbon source. The increase in the number of differentially expressed genes was such that most genes that were activated in a higher-quality carbon source did so again in one of lower quality. Thus, even in the presence of a particular carbon source of low quality, the bacterium kept genes for metabolising better nutrient sources active. This, together with the researchers' observation that poor nutrient sources led to increased motility, suggested a *foraging* strategy where the bacterium is on the lookout for a nutrient of better quality than it sees. To my knowledge, the details of how this is achieved has not been established.

5.7.5.7 Complex relationships between DNA binding and transcriptional effects of nucleoid-associated proteins

Nucleoid-associated proteins (NAPs) are a class of DNA-binding proteins, which typically bind to a large number of sites on the chromosome and modulate the three-dimensional architecture of the bound DNA. According to an early estimate, there are about 12 NAPs in the genome of *E. coli*; the exact number is a matter of debate. The interactions of NAPs with the chromosome have far-reaching effects on multiple aspects of chromosome biology, including transcription. The relationship between the binding of various NAPs to the chromosome and gene expression has been the subject of several genomic studies.

A commonly investigated NAP is the protein FIS, which is expressed primarily during early exponential phase, when it gives a boost to the expression of ribosomal operons. The expression level of this protein declines

[148] See the CRP CHIP-chip study described above.

[149] Liu M., Durfee T., Cabrera J. E., Zhao K., Jin D. J., Blattner F. R. 2005. 'Global transcriptional programs reveal a carbon source foraging strategy by *Escherichia coli*.' *Journal of Biological Chemistry* 280: 15921–15927.

sharply after the mid-exponential phase and is hardly detectable during the stationary phase. There have been at least three ChIP-chip/seq studies of FIS in *E. coli*. In the first, Grainger and colleagues performed a straightforward ChIP-chip analysis of FIS (among other NAPs) binding to the *E. coli* genome, using a low-resolution tiling microarray (~22,000 probes covering the genome).[150] They found that a majority of probes that were defined as being bound by FIS corresponded to A+T-rich non-coding regions of the genome. Later, Cho and co-workers used a higher-density genome tiling microarray (~370,000 probes; 17-fold more probes than Grainger et al.'s microarray) to perform a ChIP-chip of FIS.[151] This study found ~900 binding regions for FIS, compared to ~200 identified by Grainger's study. Cho and colleagues further constructed a FIS$^-$ strain of *E. coli*, and found that ~20% of all genes responded moderately to this deletion. However, there was little association between the location of FIS binding regions and the genes that changed in expression in FIS$^-$ compared to the wild type. Finally, Kahramanoglou and colleagues performed a ChIP-seq of FIS in the early-and mid-exponential phases of growth.[152] Over ~1,200 FIS-binding regions were identified, most of which were consistent between the two phases of growth. The studies by Cho et al. and Kahramanoglou et al. studies recovered the known FIS-binding DNA motif from the binding regions they had defined. Both also noted that as many as three-fourths of FIS-binding sites are within coding regions, in contrast to the findings of Grainger and colleagues. Given that >90% of the *E. coli* genome comprises genes, one could statistically say that FIS does prefer to bind to intergenic regions. Though the binding regions identified by Kahramanoglou agreed reasonably well with those described by Cho, there was hardly any correlation between these two studies and that by Grainger. This remains a puzzling observation. However, it is notable that the ChIP-seq results of FIS show some similarity to those observed for CRP, in that the signals do not score much higher above the background (Fig. 5.25). Kahramanoglou et al. also performed a transcriptome analysis–using standard gene expression microarrays–of a FIS$^-$ strain and compared it with the wild type (Fig. 5.26). Consistent with the findings of Cho et al., most binding sites of FIS in the Kahramanoglou study did not result in a proximal gene expression change; and most genes which changed

---

150 Grainger D. C., Hurd D., Goldberg M. D. and Busby S. J. 2006. 'Association of nucleoid proteins with coding and non-coding segments of the *Escherichia coli* genome.' *Nucleic Acids Research* 34: 4642–4652.

151 Cho B. K., Knight E. M., Barrett C. L. and Palsson B. O. 2008. 'Genome-wide analysis of Fis binding in *Escherichia coli* indicates a causative role for A-/AT-tracts.' *Genome Research* 18: 900–10.

152 Kahramanoglou C., Seshasayee A. S., Prieto A. I., Ibberson D., Schmidt S., Zimmermann J., Benes V., Fraser G. M. and Luscombe N. M. 2011. 'Direct and indirect effects of H-NS and Fis on global gene expression control in *Escherichia coli*.' *Nucleic Acids Research* 39: 2073–2091.

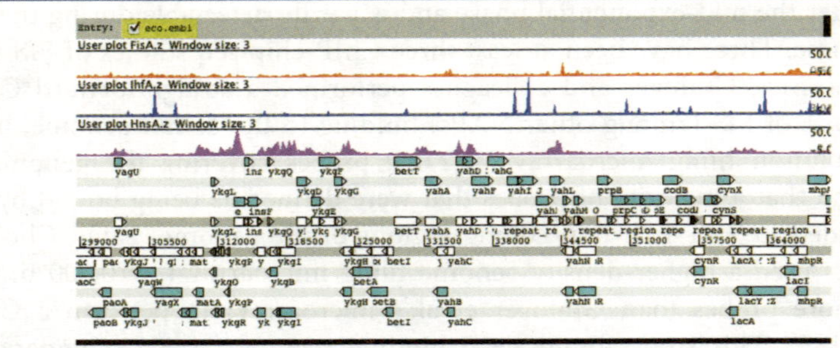

Figure 5.25 *ChIP-seq signals for three sequence-specific nucleoid-associated proteins in E. coli.* This figure shows ChIP-seq signals for three *E. coli* NAPs: H-NS (magenta), FIS (red) and IHF (blue). Whereas for IHF and (to a lesser extent) H-NS, signals score well over the background, this is less so for FIS. Also note the long binding tracts for H-NS. This figure was drawn using the Artemis Genome Browser. These data are from Kahramanoglou et al. 2011 and Prieto et al. 2012.

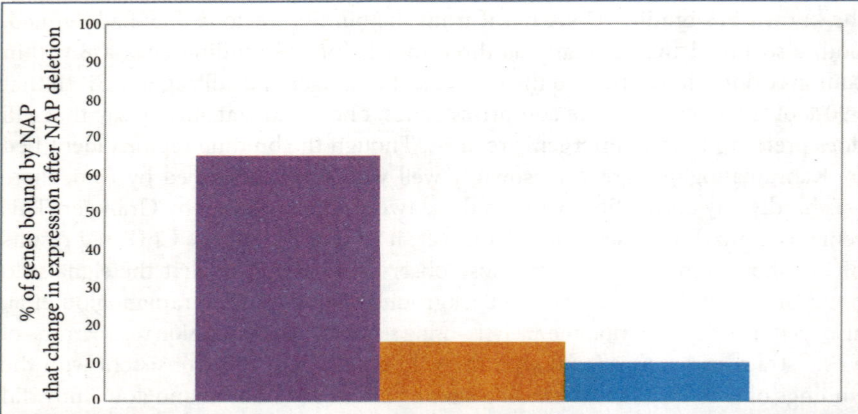

Figure 5.26 *Agreement between the binding and measurable transcriptional effects of NAPs.* This figure shows percentage of genes that bound by a NAP (H-NS, FIS or IHF), which are are also differentially expressed in a strain lacking the corresponding NAP when compared to the wild type. Whereas only a small proportion of genes bound by FIS or IHF change in expression in the corresponding NAP deletion, there is a greater agreement between the binding of H-NS and its effect on gene expression. These data are from Kahramanoglou et al. 2011 and Prieto et al. 2012.

in expression in FIS$^-$ did not have a strong FIS-binding site nearby. Kahramanoglou et al. explored this further. They classified genes that were bound by FIS into two: Those that showed a change in expression in FIS$^-$ and those that did not. They found that FIS-bound genes that also changed in expression in FIS$^-$ had multiple FIS-binding sites upstream of the gene, which also corresponded to a higher A+T-content of the binding site. Thus, there was a much stronger association between FIS-binding to multiple sites in intergenic regions and expression of the proximal gene. Next, these researchers also integrated the transcriptional regulatory network from RegulonDB, and found that a significant proportion of genes, which changed in expression in FIS$^-$ but was not bound by FIS were in fact targets of TFs which were regulated by FIS. Thus, a significant proportion of gene expression change in a FIS$^-$ strain could be explained by cascades of regulatory interactions downstream of FIS. In summary, FIS appears to bind to a large number of binding sites, most of which were located within genes, on the genome. Though there are some associations between certain properties of FIS-binding sites and the expression of nearby genes, the fact is that most of the binding sites have little apparent effect on gene expression.

A second NAP, called IHF, which is composed of two subunits A and B, also showed little correlation between its binding sites on the chromosome and the effect, on gene expression of a IHFA$^-$/IHFB$^-$ double mutant[153] (Figs 5.25 and 5.26). Unlike FIS, IHF is expressed throughout growth, with its expression level increasing in the stationary phase. There were clear differences between the ChIP-seq profiles of IHF and FIS. IHF showed low background, unlike FIS (and CRP). The IHFA$^-$/IHFB$^-$ double mutant had a relatively small effect on gene expression, when compared to the FIS$^-$ mutant. It is suspected that much of the transcriptional effects of IHF may be dependent on accessory proteins–such as those involved in regulating genes under σ54 control–whose activity might be dependent on specific environmental conditions.

The best conserved NAP among bacteria is the protein called HU. In many bacteria, it is present as a single copy protein and in others, including *E. coli*, it is present as two homologs (HupA and HupB) like IHF. In terms of sequence, HU is homologous to IHF. However, while IHF binds to DNA in a sequence-specific manner, HU binds non-specifically, albeit preferentially to distorted DNA. Prieto and colleagues sought to determine the DNA binding profile of HU on a genome-wide scale using ChIP-seq.[154] This experiment was performed independently for HupA and HupB. For sequence specific global TFs, these researchers had generally observed a per base read count distribution with a sharp peak to the left,

[153] Prieto A. I., Kahramanoglou C., Ali R. M., Fraser G. M., Seshasayee A. S. and Luscombe N. M. 2012. 'Genomic analysis of DNA binding and gene regulation by homologous nucleoid-associated proteins IHF and HU in *Escherichia coli* K12.' *Nucleic Acids Research* 40: 3524–3537.

[154] Prieto A. I., Kahramanoglou C., Ali R. M., Fraser G. M., Seshasayee A. S. and Luscombe N. M. 2012. 'Genomic analysis of DNA binding and gene regulation by homologous nucleoid-associated proteins IHF and HU in *Escherichia coli* K12.' *Nucleic Acids Research* 40: 3524–3537.

representing the background, and a long tail to the right, representing regions preferentially recovered by the pull down experiment. However, for HU, the right tail was largely absent, in a manner similar to a mock-IP experiment. This, according to the authors, was representative of the non-specific DNA binding characteristic of HU (Fig. 5.27). Nevertheless, the researchers found that the binding signal for HU was higher in A+T-rich DNA; this was interpreted as being consistent with the preferential binding of HU to distorted DNA structures, which are sometimes associated with high A+T-content. The ChIP-seq signal was similar for HupA and HupB, across all the four stages of growth investigated, indicating that they bound to similar sites on the DNA. This was despite previous evidence that HU existed predominantly in a HupA2 form during the exponential phase, with little role for the HupB subunit. However, it must be noted that the researchers observed comparable expression of both HupA and HupB throughout the growth. Further, the authors also noted that the binding signal for each subunit was independent of the presence of the other subunit, as measured by ChIP-seq of HupA in HupB$^-$ and vice-versa. However, the authors do mention that their ChIP-seq data may be unable to test whether the binding of one subunit of HU in the absence of the other was uniformly less throughout the genome. Finally, Prieto and colleagues also generated transcriptome data for HupA$^-$, HupB$^-$ and HupA$^-$/HupB$^-$ mutants of *E. coli*. In general, they found that the HupB$^-$ strain showed little change in gene expression. The HupA$^-$ strain had a more significant gene expression change, with the HupA$^-$/HupB$^-$ double mutant experiencing the most dramatic alteration in gene expression. Though there was an overlap between the set of genes differentially expressed in HupA$^-$ and that in the HupA$^-$/HupB$^-$ double mutant, each had a

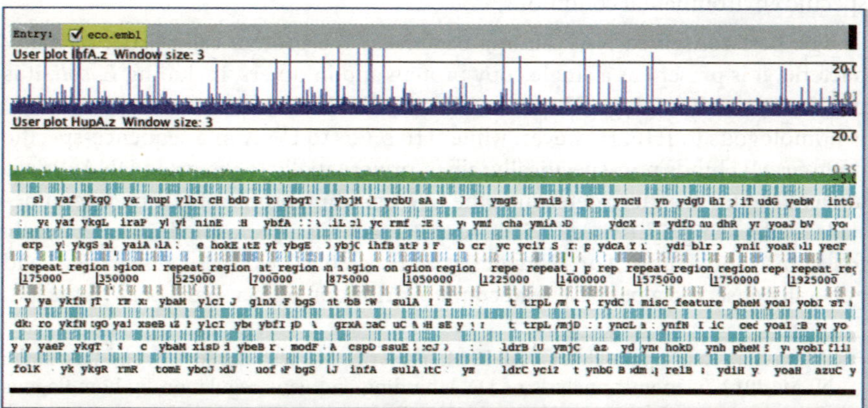

Figure 5.27 *Comparison of ChIP-seq profiles of two homologous NAPs – IHF and HU.* This figure shows tracks indicating the binding signal of IHF (blue) and HU (green) to a section of the *E. coli* chromosome. Whereas IHF shows distinct sharp peaks, HU shows a more uniform distribution across the chromosome. This figure was drawn using the Artemis Genome Browser. These data are from Prieto et al. 2012.

considerable exclusive component. Here we mention the salient properties of those genes differentially expressed in the double mutant. A significant proportion of genes responding to HupA$^-$/HupB$^-$ had previously been described as essential for growth in rich media, and were also conserved across a large number of bacterial genomes. Many of these were involved in ribosome biogenesis. These appeared to be consistent with the conservation of HU across many bacterial genomes, and the pleiotropic phenotype–including slow growth–of the double mutant. The genes were also found to have high gene expression in the wild type, and high DNA gyrase binding, as derived from an earlier ChIP-chip resource for DNA gyrase.[155] A previous study of the HupA$^-$/HupB$^-$ transcriptome–performed by Berger and colleagues from Georgi Muskhelishvili's laboratory–had shown a similar, albeit stronger, association between DNA gyrase binding and the transcriptional effects of HU.[156] In both the studies, it was observed that highly expressed ribosomal genes were in fact further up-regulated in the HupA$^-$/HupB$^-$ double mutant. Berger and colleagues also noted using fluorescence microscopy that the formation of transcriptional foci, i.e., bright spots of RNA polymerase clustered around highly expressed genes, was impaired in the HupA$^-$/HupB$^-$ double mutant. The first (to my knowledge) comprehensive transcriptome analysis of HU was performed by Oberto and colleagues,[157] who performed microarray experiments of gene expression of HupA$^-$, HupB$^-$ and HupA$^-$/HupB$^-$ strains of *E. coli* during several stages of growth. Instead of performing a standard analysis of differential expression, they performed a clustering analysis of a gene expression matrix in which each column represented a sample (strain *x* growth phase), and each row, a gene. They found that ~350 genes were regulated by HU in at least one of its forms. In agreement with the later study by Prieto and colleagues, HupB$^-$ had little effect on gene expression during the exponential phase. Unlike the studies by Prieto and Berger, Oberto and colleagues do not report a possible connection between DNA supercoiling and the transcriptome of the various HU mutants. Instead, they note that various genes involved in stress, namely acid and osmotic stress and SOS response, responded to the HU mutant, alongside various genes involved in anaerobic energy metabolism. Overall, it seems that the transcriptional effects of HU are complex, and that there may be extensive strain-specific variation in its effects,[158] despite its high conservation across bacteria.

---

155 Jeong K. S., Ahn J. and Khodursky A. B. 2004. 'Spatial patterns of transcriptional activity in the chromosome of *Escherichia coli.*' *Genome Biol.* 5(11): R86.

156 Berger M., Farcas A., Geertz M., Zhelyazkova P., Brix K., Travers A. and Muskhelishvili G. 2010. 'Coordination of genomic structure and transcription by the main bacterial nucleoid-associated protein HU.' *EMBO Reports* 11: 59–64.

157 Oberto J., Nabti S., Jooste V., Mignot H. and Rouviere-Yaniv J. 2009. 'The HU regulon is composed of genes responding to anaerobiosis, acid Stress, high Osmolarity and SOS induction.' *PLoS One* 4: e4367.

158 Each of the three studies discussed here had used a different strain of *E. coli.*

5.7.5.8 A transcriptional silencing system for horizontally-acquired genes

We had emphasised earlier in this book that horizontal gene transfer is a major means of evolution of bacteria. Besides allowing an immediate exploration of novel phenotypes and niches, a horizontally-acquired gene has more protracted effects on the host genome by imposing its own selective pressures. The establishment of a horizontally-acquired gene is a function of the selective advantage it provides to its host, and the cost it imposes in terms of potential toxicity or high gene expression. Many horizontally-acquired genes, including prophage remnants, appear to contribute to host fitness under certain conditions of stress. It is believed that perpetual high gene expression is a barrier to the establishment of a horizontally-acquired gene, especially of one whose benefit is limited to selected conditions. Therefore, it makes sense for these genes to be transcriptionally repressed under most conditions.

H-NS is a NAP, commonly present in enterobacteria. It binds to curved and/or A+T-rich DNA, on which it oligomerises. The coating of bound DNA by H-NS may lead to the formation of DNA–H-NS–DNA bridges or stiff rods, which are expected to suppress transcription. Therefore, H-NS is a repressor of the expression of genes encoded by A+T-rich and/or curved DNA. It also acts as a major organiser of the chromosome, as suggested by high resolution fluorescence microscopy, which revealed that all the binding sites of H-NS collapse into two foci inside the *E. coli* cell.[159] This might suggest that H-NS, as a transcriptional repressor, forms the *E. coli* heterochromatin.

Lucchini and co-workers from Jay Hinton's laboratory performed a ChIP-chip analysis of H-NS and the RPO in *Salmonella enterica Typhimurium* using tiling microarrays.[160] Though the researchers describe the presence of a high binding signal for H-NS across most of the chromosome, they identified ~420 high-affinity binding regions for H-NS, corresponding to ~750 genes (Fig. 5.28). Genes that were bound by H-NS showed low expression in wild type cells as measured by gene expression microarray experiments. There was almost no overlap between the lists of genes defined as bound by H-NS and those by RPO. This reinforced the notion that H-NS is a strong repressor of gene expression. Though nearly half of the genes bound by H-NS were de-repressed in a H-NS$^-$ strain, the other half was not, suggesting to the researchers the absence of possible activators for these genes under the condition tested. However, it is to be noted that, unlike in the study for FIS and IHF, the association between genes bound by H-NS and those that respond to H-NS$^-$ is significantly high. The strongest correlation between

---

159 Wang W., Li G. W., Chen C., Xie X. S. and Zhuang X. 2011. 'Chromosome organization by a nucleoid-associated protein in live bacteria.' *Science* 333: 1445–1449.

160 Lucchini S., Rowley G., Goldberg M. D., Hurd D., Harrison M. and Hinton J. C. 2006. 'H-NS mediates the silencing of laterally acquired genes in bacteria.' *PLoS Pathogens* 2: e81.

H-NS binding sites and properties of the bound DNA was A+T-content. This could also be confirmed using the list of genes that were up-regulated in H-NS$^-$. Many horizontally-acquired genes in *S. enterica*, as in most enterobacteria, are A+T-rich. Thus, H-NS emerges as a silencer of horizontally-acquired genes. Of course, A+T-content, and not horizontal acquisition, is the determinant of H-NS binding, as was evident from the absence of H-NS binding to various *S. enterica* prophages of low A+T-content. H-NS would also bind to any core genes, as long as they have high A+T-content. An independent transcriptome study by Navarre and co-workers also showed that H-NS silences A+T-rich horizontally-acquired

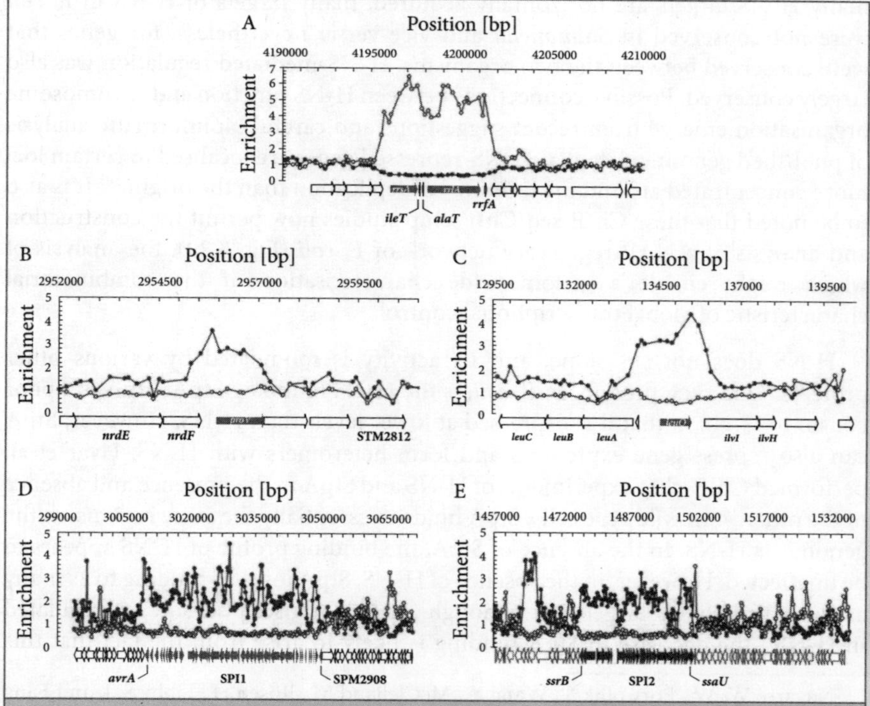

Figure 5.28 *Binding of H-NS to the Salmonella Typhimurium genome.* (A) This figure shows maps of the mutually-exclusive binding of H-NS (filled circle) and RNA polymerase (open circle) to various regions of *Salmonella enterica Typhimurium.* The two panels at the bottom show a pair of horizontally-acquired A+T-rich pathogenicity islands, which are repressed by H-NS. This figure is reproduced under the Creative Commons Attribution License from Lucchini S., Rowley G., Goldberg M. D., Hurd D., Harrison M., et al. 2006. 'H-NS mediates the silencing of laterally acquired genes in bacteria.' *PLoS Pathog* 2(8): e81. © Lucchini et al. 2006.

genes in *S. enterica.*[161] Similar findings were made from ChIP-chip studies of *E. coli* H-NS as well as by Oshima and colleagues.[162] However, this latter study, using RPO ChIP-chip, suggested that RPO was bound but locked at H-NS-bound promoters. A later ChIP-seq study of *E. coli* H-NS by Kahramanogou and co-workers reaffirmed the above findings, using additional gene expression microarrays and RPO ChIP-seq experiments. This study supported the Lucchini model that H-NS binding abrogates RPO binding to promoters. The Kahramanoglou experiment showed the presence of long-binding tracts of H-NS, consistent with its ability to oligomerise on the DNA; it was further shown that these long-binding tracts were the ones associated with gene silencing. Consistent with the notion that many H-NS targets are horizontally-acquired, many targets of H-NS in *E. coli* were not conserved in *Salmonella* and vice-versa; nevertheless, for genes that were conserved between the two organisms, H-NS-mediated regulation was also largely conserved. Possible connections between H-NS function and chromosome organisation emerge from recent suggestions and careful bioinformatic analysis of published genomic data that H-NS-repressed genes are localised to certain loci more concentrated around the terminus of replication than the origin.[163] It is also to be noted that these ChIP-seq/ChIP-chip studies now permit the construction and analysis of a NAP-regulatory network of *E. coli* (Fig. 5.29), the analysis of which might enable a genome-wide characterisation of the combinatorial characteristic of global transcriptional control.

H-NS does not act alone, and its activity is modulated by various other proteins. One such protein, in *E. coli*, is the H-NS homolog StpA. StpA is a poor cousin to H-NS in that it is expressed at lower levels than H-NS. However, StpA can also repress gene expression and form heteromers with H-NS. Uyar et al. performed ChIP-chip experiments of H-NS and StpA in the presence and absence of the other.[164] In wild type cells, StpA binds to essentially the same regions on the genome as H-NS. In the absence of StpA, the binding profile of H-NS appears to be unaffected. However, in the absence of H-NS, StpA loses its binding to as many as two-thirds of its target sites. Though StpA is transcriptionally up-regulated in H-NS^{-}, the decrease in StpA binding is likely to arise from the fact that this

161 Navarre W. W., Porwollik S., Wang Y., McClelland M., Rosen H., Libby S. J. and Fang F. C. 2006. 'Selective silencing of foreign DNA with low GC content by the H-NS protein in Salmonella.' *Science* 313: 236–38.

162 Oshima T., Ishikawa S., Kurokawa K., Aiba H. and Ogasawara N. 2006. '*Escherichia coli* histone-like protein H-NS preferentially binds to horizontally acquired DNA in association with RNA polymerase.' *DNA Research*. 13: 141–53.

163 Zarei M., Sclavi B. and Cosentino Lagomarsino M. 2013. 'Gene silencing and large-scale domain structure of the *E. coli* genome.' *Molecular Biosystems* 9: 758–67.

164 Uyar E., Kurokawa K., Yoshimura M., Ishikawa S., Ogasawara N., Oshima T. 2008. 'Differential binding profiles of StpA in wild-type and *h-ns* mutant cells: A comparative analysis of cooperative partners by chromatin immunoprecipitation-microarray analysis.' *Journal of Bacteriology* 191: 2388–2391.

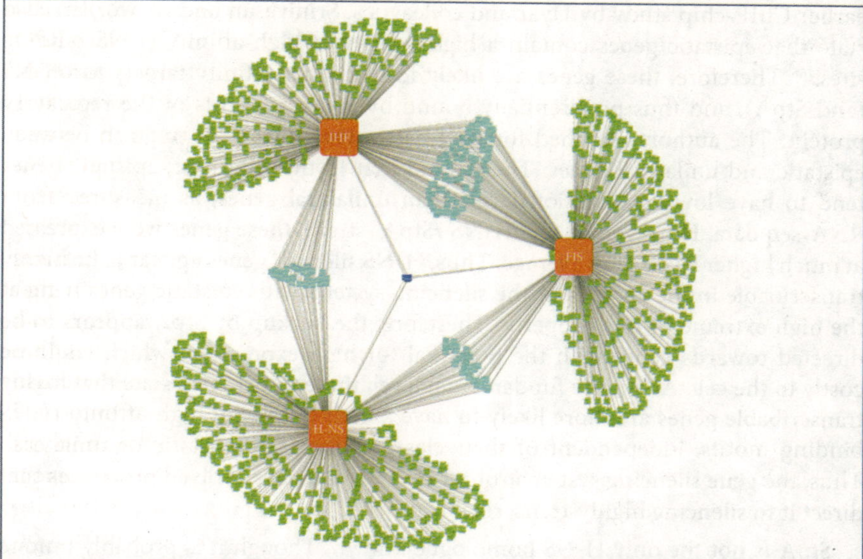

Figure 5.29 *A NAP regulatory network of E. coli.* This figure shows a network representation of genes which are bound by three sequence-specific NAPs in *E. coli* during exponential growth. Very few genes (blue) are bound by all the three NAPs; several (cyan) are bound by two of the three NAPs – FIS and IHF in particular; and most (green) bound by a single NAP. This was drawn using the Cytoscape network-rendering software. These data are from Kahramanoglou et al. 2011 and Prieto et al. 2012.

protein becomes susceptible to proteolysis in the absence of H-NS. Thus, StpA is a partial backup for H-NS. This backup could have important functions, as revealed by the findings that a H-NS$^-$/StpA$^-$ double deletion has a severe growth defect not explained by the single deletions taken together. To investigate the nature of this *epistatic* interaction between H-NS and StpA, Srinivasan and colleagues performed a transcriptome analysis, by RNA-seq, of the two single mutants and the double mutant, alongside the wild type.[165] They found that about a third of all genes which respond to a H-NS$^-$ deletion are further affected by the H-NS$^-$/StpA$^-$ double deletion; these they refer to as *epistatic* genes. The remaining two-thirds of H-NS-regulated genes, termed *unilateral*, are not affected by StpA$^-$. It is to be noted that StpA$^-$ has a global effect on gene expression only in an H-NS$^-$ background, and not in the wild type background. The differences between epistatic and unilateral genes could be explained by StpA binding data from the

[165] Srinivasan R., Chandraprakash D., Krishnamurthi R., Singh P., Scolari V. F., Krishna S. and Seshasayee A. S. 2013. 'Genomic analysis reveals epistatic silencing of expensive genes in *Escherichia coli* K-12.' *Molecular Biosystems* 9: 2021–2033.

earlier ChIP-chip study by Uyar and colleagues. Srinivasan and co-workers also state that epistatic genes contain a high density of high-affinity H-NS binding sites.[166] Therefore, these genes are likely to be higher-affinity targets for H-NS (and StpA), and thus preferentially bound by lower amounts of the regulatory protein. The authors searched for parameters that further distinguish between epistatic and unilateral genes. They found that in the wild type, epistatic genes tend to have lower expression levels than unilateral genes, as measured from RNA-seq data. However, in the H-NS$^-$/StpA$^-$ strain, these genes were expressed at much higher levels than average. Thus, H-NS silenced genes appear to be highly transcribable in the absence of the silencing system, with epistatic genes lying at the high extreme of this property. Therefore, the backup by StpA appears to be directed towards genes with the potential for high expression, which could be costly to the cell. At a more fundamental level, these researchers state that highly transcribable genes are more likely to have a high density of high-affinity H-NS binding motifs, independent of their classification into epistatic or unilateral. Thus, the gene silencing system around H-NS could have evolved properties that direct it to silencing highly-transcribable genes.[167]

StpA is not the only H-NS homologue known. Though it is probably unique in being a chromosomally-encoded H-NS homologue, other plasmid-encoded homologues of H-NS are known in at least *Salmonella* and *Shigella*. These proteins are believed to act as stealth proteins that silence gene expression from plasmid-encoded genes, while ensuring that chromosomally-encoded H-NS is not titrated out, thus affecting H-NS–chromosome interactions. One such plasmid-encoded stealth protein is called Sfh. Dillon and colleagues from the laboratories of Jay Hinton and Charles Dorman investigated the binding profile of Sfh to the chromosome of *S. enterica Typhimurium* using ChIP-chip.[168] They identified ~250 genes as bound by Sfh. Remarkably, every gene bound by Sfh was also bound by H-NS, showing that Sfh binds to a subset of H-NS-bound genes. However,

---

[166] This was based on *in-vitro* high-throughput data from Gordon and colleagues, which measured the relative affinity of H-NS to ~40,000 eight-mer nucleotide sequences. Gordon B. R., Li Y., Cote A., Weirauch M. T., Ding P., Hughes T. R., Navarre W. W., Xia B. and Liu J. 2011. 'Structural basis for recognition of AT-rich DNA by unrelated xenogeneic silencing proteins.' *Proceedings of the National Academy of Sciences USA* 108: 10690–10695.

[167] We note here that the high transcribability of these loci most likely emerge from pervasive transcription arising from many promoter-like elements present in these A+T-rich sequences. See Singh S. S., Singh N., Bonocora R. P., Fitzgerald D. M., Wade J. T. and Grainger D. C. 2014. 'Widespread suppression of intragenic transcription initiation by H-NS.' *Genes and Development* 28: 214–19.

[168] Dillon S. C., Cameron A. D., Hokamp K., Lucchini S., Hinton J. C. and Dorman C. J. 2010. 'Genome-wide analysis of the H-NS and Sfh regulatory networks in Salmonella Typhimurium identifies a plasmid-encoded transcription silencing mechanism.' *Molecular Microbiology* 76: 1250–1265.

only a small fraction of these genes were differentially expressed in Sfh$^-$, which did result in a change in expression level of over 400 genes. In H-NS$^-$, Sfh bound to more sites on the chromosome, which would otherwise have been bound by H-NS. Thus, Sfh binding to the *Salmonella* genome contrasts with that of StpA in *E. coli.* Though Sfh does bind to H-NS targets, Dillon et al. suggest that the two proteins recognise different sequence features: For example, regions that Sfh binds to–in the presence or absence of H-NS–tend to have a lower A+T-content than those that are uniquely bound by H-NS. In summary, the stealth protein Sfh 'tops-up' the cellular pool of H-NS-like proteins, minimising any disturbance that the entry of a plasmid might cause to H-NS-mediated control of chromosomal genes.

In addition are also seemingly unrelated TFs such as LeuO, which also bind to a large number of H-NS-bound sites, as determined by ChIP-chip[169] as well as *in-vitro* SELEX experiments,[170] but act as H-NS antagonists. We do not discuss these results in more detail here, except to mention that these proteins are candidates for ensuring that H-NS-mediated gene silencing is relieved when the expression of the target gene is required; this phenomenon is referred to as *anti-silencing*.

We had mentioned that though many H-NS targets are horizontally-acquired genes, many also have core cellular functions. In fact, the study by Srinivasan and colleagues had mentioned that many genes that were up-regulated in H-NS$^-$/StpA$^-$ were targets of the global TF FNR, which regulates many anaerobic energy metabolism genes. A recent genomic study by Myers and colleagues found that many sites computationally predicted to be FNR targets, were not bound by the protein as measured by ChIP-seq.[171] By integrating their FNR ChIP-seq data with similar data for NAPs, they found that as many as 90% of predicted-but-unbound FNR target motifs were bound by at least one of FIS, IHF and H-NS. Remarkably, many of these sites became occuped by FNR in a H-NS$^-$/StpA$^-$ strain. This shows that NAPs influence the function of classical TFs on a genomic scale.

We will close this section by describing an intriguing interplay between transcriptional silencing by H-NS and transcription termination by the protein Rho. In bacteria, transcription can terminate independently of any protein factors at certain sequences where mRNA forms a certain type of stem loop. In other genes, termination is caused by a complex protein called Rho. Deletion of Rho

169 Dillon S. C., Espinosa E., Hokamp K., Ussery D. W., Casadesús J. and Dorman C. J. 2012. 'LeuO is a global regulator of gene expression in Salmonella enterica serovar Typhimurium.' *Molecular Microbiology* 85: 1072–1089.

170 Shimada T., Bridier A., Briandet R. and Ishihama A. 2011. 'Novel roles of LeuO in transcription regulation of *E. coli* genome: Antagonistic interplay with the universal silencer H-NS.' *Molecular Microbiology* 82: 378–97.

171 Myers K. S., Yan H., Ong I. M., Chung D., Liang K., Tran F., Keleş S., Landick R. and Kiley P. J. 2013. 'Genome-scale Analysis of *Escherichia coli* FNR reveals complex features of transcription factor binding.' *PLoS Genetics* 9: e1003565.

is lethal, therefore, much research on this protein has either used point mutants with impaired function, or a small-molecule called bicyclomycin (BCM), which is widely believed to be a specific inhibitor of Rho. Cardinale and colleagues performed a transcriptome analysis, using gene expression microarrays, of *E. coli* treated with BCM.[172] These researchers found that BCM de-represses the expression of horizontally-acquired genes. A similar pattern was observed in certain mutants of other genes associated with Rho, suggesting that this is probably not a possible Rho-independent consequence of BCM. Though the researchers do point to the similarity this shares with the function of H-NS, this angle was not taken forward. A classical genetic study by Saxena and Gowrishankar suggested that H-NS might influence Rho action by possibly slowing down the elongating RPO such that it becomes more susceptible to termination by Rho.[173] A recent study, using tiling microarrays and deep sequencing, compared the transcriptome of BCM treated cells and the binding profile of H-NS in non-treated cells.[174] This study explicitly showed that ~80% of genes that are transcriptionally up-regulated by BCM treatment are also H-NS targets. This result was interpreted based on the Saxena and Gowrishankar model. Very recently, Chandraprakash and Seshasayee performed a ChIP-seq study of *E. coli* H-NS in the presence and absence of BCM.[175] They found that BCM treatment globally decreases the binding signal of H-NS across its range of targets. Thus, this study presents evidence that proper Rho function may be necessary for effective H-NS–DNA interaction and transcriptional suppression. Though the mechanism behind this is not known, the two opposing models are not incompatible; instead, a positive feedback loop between H-NS and Rho might, in fact, help reinforce gene silencing.

5.7.6 Transcriptional control by the small-molecule alarmone ppGpp

We had discussed the role of σ-factor switching in the complex transcriptional response of *E. coli* to starvation, among other stresses. An important player in this transition is the nucleotide ppGpp (guanosine tetra-phosphate). During amino acid starvation, the loading of uncharged tRNAs on the ribosome is thought to

[172] Cardinale C. J., Washburn R. S., Tadigotla V. R., Brown L. M., Gottesman M. E. and Nudler E. 2008. 'Termination factor Rho and its cofactors NusA and NusG silence foreign DNA in *E. coli*.' *Science* 320: 935–38.

[173] Saxena and Gowrishankar. 2011. 'Modulation of Rho-dependent transcription termination in *Escherichia coli* by the H-NS family of proteins.' *Journal of Bacteriology* 193: 3832–3841.

[174] Peters J. M., Mooney R. A., Grass J. A., Jessen E. D., Tran F. and Landick R. 2012. 'Rho and NusG suppress pervasive antisense transcription in *Escherichia coli*.' *Genes and Development* 26: 2621–2633.

[175] Chandraprakash D. and Seshasayee A. S. 2014. 'Inhibition of factor-dependent transcription termination in *Escherichia coli* might relieve xenogene silencing by abrogating H-NS-DNA interactions *in-vivo*.' *Journal of Biosciences* 39, 1: 53–61.

activate the protein called RelA, which synthesises ppGpp. ppGpp binds directly to the RPO and prevents transcription, most notably of stable RNA. This in turn decreases the levels of players involved in protein synthesis. In addition, the cessation of transcription from the predominant stable RNA is also believed to free up RPO for transcribing other stress-response-related genes. ppGpp can also directly stimulate transcription of certain metabolic genes involved in amino acid biosynthesis.

Traxler and colleagues performed a study investigating the transcriptional effects of ppGpp, in standard growth conditions as well as under startvation of the amino acid isoleucine, on *E. coli.*[176] They grew the wild type and an otherwise isogenic ppGpp^{-}[177] strain of *E. coli* in minimal medium supplemented with amino acids, and in a second medium, that contained a much lower concentration of one amino acid isoleucine. In the isoleucine-limited medium the bacteria grew well, albeit after a long lag phase. The researchers then used microarrays to measure gene expression of the wild type and the ppGpp^{-} mutant grown till entry into stationary phase, which was called by Traxler and colleagues as the *growth-arrested* stage. Control experiments were performed for the bacteria growing rapidly in the exponential phase in media containing isoleucine. Both the wild type and the ppGpp^{-} strains showed extensive transcriptional response to isoleucine starvation, with ~1,000 genes being differentially expressed. The number of genes down-regulated by isoleucine starvation was not very different from that up-regulated. In the wild type cells, many metabolic enzymes and genes involved in general stress response were up-regulated. Among the genes induced in the wild type by isoleucine starvation were those involved in central metabolism, possibly leading to an increased flux into pyruvate, and genes involved in the synthesis of branched chain amino acids and their precursors. As expected, many genes involved in ribosome structure and translation were down-regulated. Though the ppGpp^{-} strain also showed an extensive transcriptional response to isoleucine deprivation, the set of genes differentially expressed was very different. For example, the down-regulation of ribosomal genes–observed in the wild type–was absent; as was the induction of genes involved in amino acid biosynthesis. Interestingly, and in contrast to the wild type, many genes from central metabolism were down-regulated in the ppGpp^{-} mutant. The ppGpp^{-} strain appeared to be larger, with higher biomass. Consistent with this, macromolecular biosynthesis–including genes involved in cell division and DNA replication–was generally up-regulated in this mutant, when compared to the wild type. Thus, this study showed that the involvement of ppGpp in regulating transcription in response to amino acid

---

176 Traxler M. F1., Summers S. M., Nguyen H. T., Zacharia V. M., Hightower G. A., Smith J. T. and Conway T. 2008. 'The global, ppGpp-mediated stringent response to amino acid starvation in *Escherichia coli*.' *Molecular Microbiology* 68: 1128–1148.

177 The ppGpp^{-} strain lacked *relA* and *spoT*. SpoT is a second ppGpp synthase in *E. coli*. Its activity is less robust than that of RelA. SpoT also has ppGpp hydrolase activity.

starvation was not limited to amino acid biosynthesis and translation, but in general to macromolecular biosynthesis and central metabolism.

5.7.7 RNA chaperones and their regulons

The importance of small RNAs (sRNA) as regulators of gene expression in bacteria–if not during transcription initiation, at least at RNA stability and translation–is becoming increasingly appreciated. Many known sRNA regulators of gene expression associate with target mRNAs through interrupted base-paired stretches. These sRNAs require a chaperone protein called Hfq to find and bind to their target mRNAs. Hfq is a highly-conserved RNA-binding protein, with specificity to unstructured, A+U-rich stretches. It is definitely a global regulator of gene expression, with its knockout resulting in pleiotropic phenotypes of *E. coli*.

We had described how genome-wide experimental annotation protocols have helped catalog sRNAs in several bacteria. In this section, we will discuss how genomic studies have identified RNA molecules that bind to Hfq. In a pioneering study, Zhang et al. from the laboratories of Gisela Storz and Susan Gottesman performed an RNA-binding-protein equivalent of a ChIP-chip experiment.[178] In such a RIP-chip experiment, RNA molecules that are bound to a protein of interest are isolated after cross-linking and immunoprecipitation against the protein of interest, and then subjected to microarray hybridisation. Zhang and colleagues performed this experiment for Hfq in *E. coli* grown to mid-or stationary phase in LB or in minimal glucose media. The RNA was hybridised to a microarray containing 15-25-mer probes for each ORF and probes for most intergenic regions. Using arbitrary microarray intensity thresholds, the researchers classified groups of probes into various levels of confidence for being bound to Hfq. Of the 12 previously known Hfq-associated sRNAs, Zhang et al. could assign the highest level of confidence to 11, showing that their approach was reliable. At least 15 of the 46 sRNAs known at the time of the publication were shown to bind to Hfq. A caveat that the researchers do describe is the false identification of some highly-expressed sRNAs as bound to Hfq. Using data from probes belonging to intergenic regions, the authors identified at least 20 novel candidate Hfq-bound sRNAs. Many mRNAs were also found to be bound to Hfq. In at least one complex ribosomal operon with known post-transcriptional processing, Hfq binding was suggested to inhibit its degradation. Finally, Hfq was shown to bind to at least one tRNA operon, where it might compete with RNA nucleases and protect the RNA from degradation. Thus, the RIP-chip approach enabled the detection of several novel Hfq-bound RNAs, leading to the delineation of novel functions for the protein.

---

178 Zhang A., Wassarman K. M., Rosenow C., Tjaden B. C., Storz G. and Gottesman S. 2003. 'Global analysis of small RNA and mRNA targets of Hfq.' *Molecular Microbiology* 50: 1111–1124.

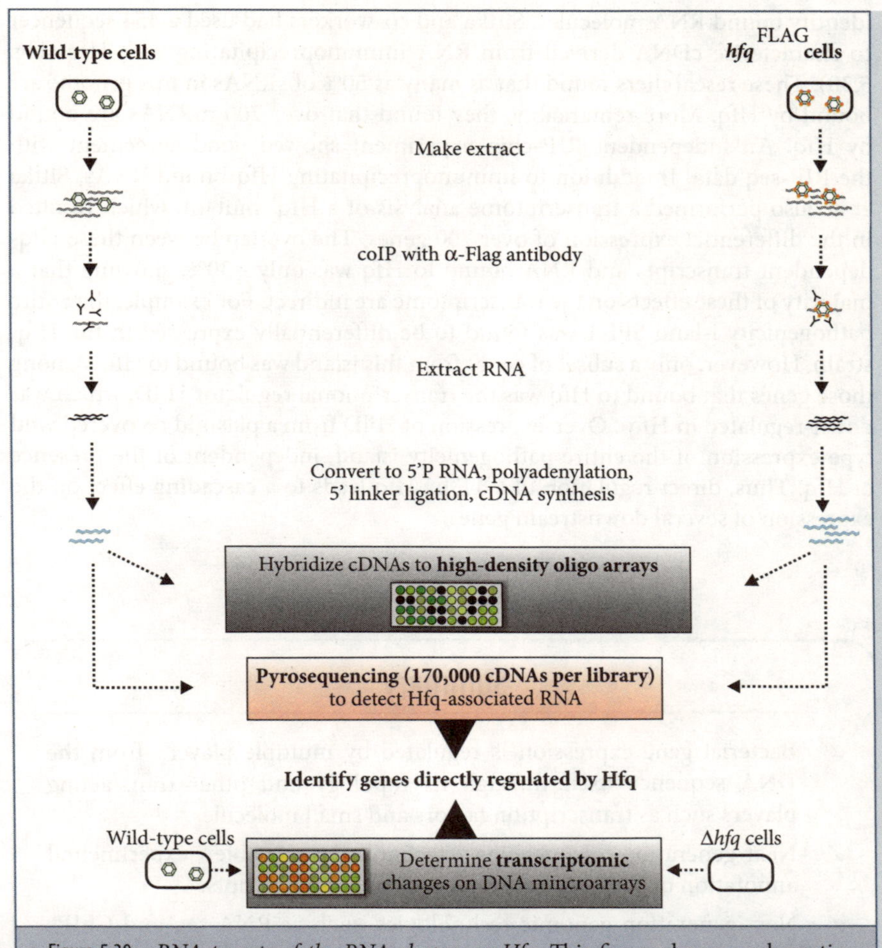

Figure 5.30 *RNA targets of the RNA chaperone Hfq.* This figure shows a schematic representation of a procedure for identifying targets of the RNA chaperone Hfq. This figure is reproduced under the Creative Commons Attribution License from Sittka A., Lucchini S., Papenfort K., Sharma C. M., Rolle K., et al. 2008. 'Deep sequencing analysis of small noncoding RNA and mRNA targets of the global post-transcriptional regulator Hfq.' *PLoS Genet* 4, 8: e1000163. © Sittka et al. 2008.

In a later study, Sittka and colleagues from the laboratories of Jorg Vogel and Jay Hinton, performed a RIP-seq experiment on Hfq in *S. enterica Typhimurium*.[179] This, similar to ChIP-seq, differs from RIP-chip in its use of deep sequencing to

179 Sittka A., Lucchini S., Papenfort K., Sharma C. M., Rolle K., Binnewies T. T., Hinton J. C. and Vogel J. 2008. 'Deep sequencing analysis of small noncoding RNA and mRNA targets of the global post-transcriptional regulator Hfq.' *PLoS Genetics* 4: e1000163.

identify bound RNA molecules. Sittka and co-workers had used a 454 sequencer to characterise cDNA derived from RNA immunoprecipitating with Hfq (Fig. 5.30). These researchers found that as many as 50% of sRNAs in this genome are bound by Hfq. More remarkably, they found that over 700 mRNAs are bound by Hfq. An independent RIP-chip experiment showed good agreement with the RIP-seq data. In addition to immunoprecipitating Hfq-bound RNAs, Sittka et al. also performed a transcriptome analysis of a Hfq$^-$ mutant, which resulted in the differential expression of over 700 genes. The overlap between these Hfq-dependent transcripts and RNA bound to Hfq was only ~30%, showing that a majority of these effects on the transcriptome are indirect. For example, the entire pathogenicity island SPI-1 was found to be differentially expressed in the Hfq$^-$ strain. However, only a subset of genes from this island was bound to Hfq. Among those genes that bound to Hfq was the transcriptional regulator HilD, which was down-regulated in Hfq$^-$. Over-expression of HilD from a plasmid recovered wild type expression of the entire pathogenicity island, independent of the presence of Hfq. Thus, direct regulation of a TF by Hfq leads to a cascading effect on the expression of several downstream genes.

Summary

- ✓ Bacterial gene expression is regulated by multiple players–from the DNA sequence itself through its topology and other trans-acting players such as transcription factors and small molecules.
- ✓ Next-generation sequencing technologies enable experimental annotation of bacterial genomes by defining transcripts.
- ✓ Next-generation genomic technologies–such as RNA-seq and ChIP-seq–can be used to construct transcriptional regulatory networks, involving in principle all types of known regulators.
- ✓ Comparative genomic approaches, combined with next-generation sequencing technologies, allow analysis of evolution of regulatory structures across organisms.
- ✓ Fluorescence reporter-based approaches allow dissection of regulatory circuits at high temporal resolution.

DNA Methylation in Bacteria: A Case for Bacterial Epigenetics

6

6.1 Introduction

In the previous chapters, we had discussed genome-scale studies of bacterial adaptation by genome content and gene expression control. In these mechanisms, there is either a change in the genotype of the bacterium; or there is a change in the protein occupancy of the genome, which is not necessarily heritable. A third mechanism of establishing identity–best studied in complex eukaryotes–is by *epigenetics*. The word essentially means 'in addition to genes'. This, in the broadest sense used by biologists today, refers to–for example–changes in gene expression which are heritable, but do not include a change in genotype. These are in contrast to the phenomena discussed in the previous chapters: Comparative genomic approaches reveal changes in genetic content, whereas mechanisms of regulation of gene expression as discussed earlier are not necessarily heritable.

A classical epigenetic mechanism involves methylation of DNA–namely adenine and cytosine bases. Methylation of cytosine at what are known as CpG islands is a reasonably well-studied gene regulatory mechanism in higher eukaryotes. Phenomena such as the specificity of certain DNA methylating enzymes to hemi-methylated DNA, as well as the competition between DNA methylating enzymes and other DNA-binding proteins, ensure that certain–if not all–DNA methylation patterns are inherited. In bacteria, the interplay of DNA methylation and DNA-binding transcription factors can cause phase variation, with distinct sub-populations exhibiting heritably different gene expression states; this is distinct from the genetic mechanism of phase variation at contingency loci in the genomes of bacteria such as *Helicobacter pylori*, discussed earlier in this book. The current short chapter does not deal extensively with the intricate biology of bacterial DNA methylation and epigenetics; for this the readers are referred to authoritative recent reviews by Casadesus and colleagues.[1] Instead, we briefly discuss the comparative genomics of enzymatic systems involved in DNA methylation and novel genome-scale approaches for detecting DNA methylation at a single-base resolution.

6.2 DNA methyltransferases in bacteria: From restriction–modification systems

DNA methyltransferases in bacteria are arguably best-known in the context of restriction–modification (R–M) systems. R–M systems comprise two enzymatic activities (Fig. 6.1). The 'restriction' component is an endonuclease, which cleaves DNA at specific sequences. The 'modification' component is the DNA methyltransferase, which adds a methyl group to either a cytosine or an adenine within the same sequence motif targeted for cleavage by its cognate restriction enzyme. DNA methylation alters the activity of the restriction endonuclease at that site: Most restriction endonucleases cleave unmethylated DNA leaving the methylated DNA intact. Therefore, DNA methylation protects DNA from being cleaved by its cognate restriction endonuclease.

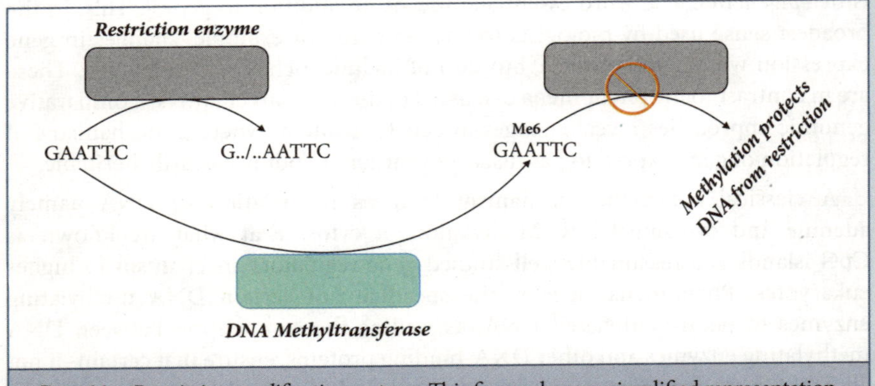

Figure 6.1 *Restriction modification systems.* This figure shows a simplified representation of the activity of restriction-modification systems, where DNA methylation protects a stretch of DNA from cleavage by the cognate restriction endonuclease.

---

1 (a) Casadesus and Low. 2006. 'Epigenetic gene regulation in the bacterial world.' *Microbiology and Molecular Biology Reviews* 70: 830–56; (b) Wion and Casadesus. 2006. 'N6-methyladenine: An epigenetic signal for DNA-protein interactions.' *Nature Reviews Microbiology* 4: 183–92.

An R–M system acts as an immune mechanism for bacteria against invading DNA, including bacteriophages. The restriction endonuclease would rapidly cleave the incoming DNA as long as its target sites are unmethylated, leaving the methylated host genome untouched. However, many R–M systems are plasmid-encoded and therefore products of recent horizontal transfer. Though protection from invading bacteriophages may be a selective force in favour of maintaining these acquired R–M systems, possible holes in this form of defence would be the tight target specificity of these enzymes, as well as the ability of bacteriophages to evolve their own mechanisms of DNA methylation. Further, R–M systems exhibit selfish behaviour in their bacterial hosts. Cells, which have lost plasmids encoding the R–M systems, undergo a form of post-segregational killing. This is because, as cell division occurs following the loss of the R–M-encoding plasmid, the protein levels of the two enzymes decrease to such a level that the methyltransferase levels are not sufficient to protect all target sites from cleavage by the restriction enzyme. Thus, a bacterium, once endowed with an R–M system, is under pressure to maintain it, unless it can selectively lose the gene encoding the restriction enzyme component. Such selective loss of the restriction enzyme leaves behind what is termed a *solitary* or an *orphan* DNA methyltransferase. Some solitary DNA methyltransferases, such as Dam in *E. coli* and CcrM in *C. crescentus*, have been well-characterised for their central regulatory roles influencing replication and cell division, and transcription. Besides offering such conventional beneficial functions, solitary methyltransferases can also provide immunity against invasion by a parasitic R–M system of the same target sequence specificity.

In a recent computational comparative genomic study, Seshasayee and colleagues sought to study the occurrence of DNA methyltransferases as components of R–M systems or as solitary enzymes.[2] To pursue this study, the researchers used a set of ~200 reference R–M system sequences from the REBASE database and searched for their homologues across >1,000 fully-sequenced prokaryotic genomes. Methyltransferase sequences thus obtained were classified as R–M components or solitary depending on the presence of a neighbouring gene encoding a restriction enzyme. The researchers found that over 75% of all the methyltransferases they discovered were solitary. Thus, methyltransferases are more likely to be found as solitary enzymes rather than R–M components. By searching for orthologs of each methyltransferase sequence within genera, the researchers found that R–M methyltransferases were poorly conserved, whereas solitary methyltrasnferases were nearly as well conserved as any average gene. Certain organisms, such as *Helicobacter pylori*, which contained conserved R–M systems also tended to code for multiple R–M systems with potentially distinct target sequences. The fact that these organisms were naturally transformable suggested to the researchers that defence against incoming DNA was a strong

---

2 Seshasayee A. S., Singh P. and Krishna S. 2012. 'Context-dependent conservation of DNA methyltransferases in bacteria.' *Nucleic Acids Research* 40: 7066–7073.

enough selective force for the maintenance of R–M systems in these organisms. Even in this species, the occurrence and activities of many R–M systems differ between strains.[3] Further, the researchers were able to detect signatures of horizontal transfer for both R–M and solitary methyltransferases. Finally, the researchers found that a significant proportion of orphan methyltransferases were encoded adjacent to a gene with relatively weak sequence similarity to restriction enzymes. This was interpreted by the authors as genome-scale evidence that solitary methyltransferases were derived by the selective 'degradation' of the gene encoding the restriction enzyme in R–M systems (Fig. 6.2). An earlier computational study by Rocha and co-workers had found that many bacterial genomes avoid short palindromic sequences, and that this could emerge from selection against the lethal pressure imposed by R–M systems.[4] Complementing this finding, Seshasayee and colleagues found that the constraint against the predicted target site of an R–M system was stronger for genomes encoding a conserved R–M system; genomes encoding recently-acquired R–M systems showed less selection against its predicted target sequence; finally, there was little selection against the target sites of solitary methyltransferases. Thus, many solitary methyltransferases, some of which could have core cellular functions, are probably evolved from R–M systems, and the presence of R–M systems and/or a corresponding solitary methyltransferase could have distinct effects on genome composition.

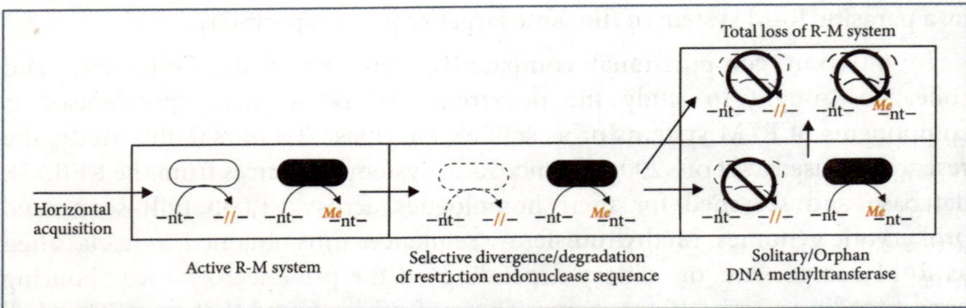

Figure 6.2 *A model for the evolution of DNA methyltransferases in bacteria.* DNA methyltransferase genes are typically acquired as part of restriction-modification systems by horizontal gene transfer. Selective loss of the endonuclease component can lead to solitary DNA methyltransferases. This model is based on Lin et al. 2001 and Seshasayee et al. 2012.

3 Lin L. F., Posfai J., Roberts R. J. and Kong H. 2001. 'Comparative genomics of the restriction-modification systems in Helicobacter pylori.' *Proceedings of the National Academy of Sciences USA* 98: 2740–2745.

4 Rocha E. P., Danchin A. and Viari A. 2001. 'Evolutionary role of restriction/modification systems as revealed by comparative genome analysis.' *Genome Research* 11: 946–58.

6.3 Identifying sites of DNA methylation on a genomic scale

6.3.1 Methylated DNA immunoprecipitation

An important contribution of genomics to the field of epigenetics is in the identification of sites of DNA methylation. Much research in this direction has been driven by the recognition of the important roles of DNA cytosine methylation to various developmental and disease processes in higher eukaryotes. As a result, progress in detecting cytosine methylation has been more prominent than that for adenine methylation, which nevertheless is an important epigenetic mark in bacteria. Two classes of genome-scale methods have been applied to the detection of DNA methylation. The first method uses antibodies specific for the methylated bases to perform a ChIP-like experiment, followed by the analysis of the immunoprecipitated DNA by microarrays or deep sequencing. This technique, dubbed MeDIP (for Methylated DNA Immuno Precipitation), has been extensively adopted for detecting methylated cytosines, though this in principle should be applicable to the detection of methylated adenines also subject to the availability of the right antibodies. The MeDIP approach however suffers from two drawbacks. The first is that it does not immediately point to the specific base position that is methylated; instead, it identifies broader loci containing one or more methylated bases. Secondly, the experiment may be more likely to pull down loci containing a cluster of methylated bases rather than those with a single (or a few) methylated base(s).

6.3.2 Bisulphite sequencing

Both the above limitations of the MeDIP technique are circumvented by the use of what is called bisulphite sequencing, which is applicable to the detection of methylated cytosines but not adenines. In this technique, DNA is first treated with a reagent such as sodium bisulphite, which converts unmethylated cytosines to uracils (later converted to thymines by PCR), but leaves methylated cytosines unchanged. If such bisulphite-treated DNA is sequenced, all unmethylated cytosines will be read as thymines and not as cytosines, thus permitting a distinction between methylated and unmethylated cytosines.

Before we move further and discuss a study applying bisulphite sequencing–and to a lesser extent MeDIP–to the analysis of cytosine methylation in *E. coli*, we will briefly review the computational challenges faced in the mapping of bisulphite sequencing reads to the reference genome, an obviously essential step in identifying methylated cytosines. The primary reason why the mapping of bisulphite sequencing reads to the reference genome is a different problem to other applications is the lower complexity of the converted sequence when compared to the reference. Most cytosines are unmethylated and are therefore read as thymines after bisulphite treatment. In addition, the opposite strand of DNA–if generated by PCR following bisulphite treatment will similarly have a large number of guanines converted to adenines. Therefore, sequencing reads from a bisulphite experiment could look very different from the reference genomic DNA sequence, thus presenting novel computational challenges.

Algorithms for mapping bisulphite sequencing reads to a reference genome first convert all cytosines to thymines in both the read and the reference genome sequences (Fig. 6.3). To consider the possibility that the read might have arisen

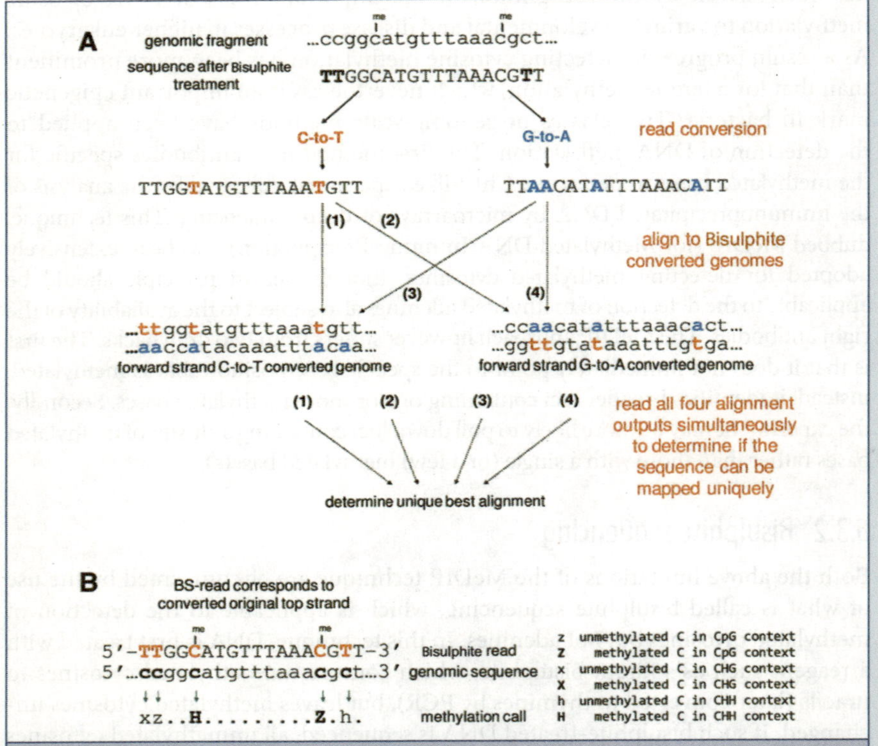

Figure 6.3 *A schematic representation of deriving methylation calls from bisulphite sequencing.* This figure shows the software Bismark's approach to bisulphite mapping and methylation calling. (A) Reads from a BS-Seq experiment are converted into a C-to-T and a G-to-A version and are then aligned to equivalently converted versions of the reference genome. A unique best alignment is then determined from the four parallel alignment processes [in this example, the best alignment has no mismatches and comes from thread (1)]. (B) The methylation state of positions involving cytosines is determined by comparing the read sequence with the corresponding genomic sequence. Depending on the strand a read mapped against this can involve looking for C-to-T (as shown here) or G-to-A substitutions. This illustration and the legend are reproduced with permission from Krueger F., Andrews S. R. 2011. 'Bismark: A flexible aligner and methylation caller for Bisulphite-Seq applications.' *Bioinformatics* Jun 1; 27, 11: 1571–1572.

from the strand opposite to that with the methylated base, guanine to adenine-converted versions of the read and the genome are also generated. Thus, differences between the reference and the read sequence caused by the bisulphite treatment are masked. Now, four alignments between the C->T or G->A converted read and the reference sequence are possible. The best alignment provides the genomic coordinate to which a read maps. When the original read sequence is compared to the aligned portion of the unconverted reference genome, all unmethylated cytosines will be seen as thymines in the read, whereas methylated cytosines will remain as cytosines. Similarly for guanosines and adenines in the opposite strand. This now enables a call on the methylation status of every cytosine on the genome. Since a methylation call is made for each read that maps to a cytosine, it is possible to measure the proportion of reads which show methylation at a given site. This number, which is generally termed the extent of methylation, may be reflective of the proportion of the genomic DNA molecules in the sample with a methylated cytosine at that position. Several readily-available open source tools enable mapping of bisulphite reads to a reference genome. These include software such as BSMAP[5] and BS Seeker.[6] Other software such as BISMARK[7] perform mapping as well as make methylation calls.

6.3.3 DNA cytosine methylation in laboratory *E. coli*

In a recent study, Kahramanoglou and colleagues performed a genome-wide analysis of cytosine methylation in *E. coli* (K12 MG1655).[8] These researchers treated *E. coli* genomic DNA–isolated at different stages of a batch culture in rich medium–with a commercially available bisulphite reagent. The resulting DNA, after PCR amplification, was subjected to Illumina sequencing, producing ~30-mer reads after the removal of multiplexing barcodes. These sequence reads were mapped back to the reference genome using BISMARK, and the extent of methylation (termed *Rmet* by the researchers) calculated for each cytosine at each of the four stages of growth investigated in this study. The only known DNA cytosine methyltransferase in the organism studied was the solitary methyltransferase Dcm, which methylates the internal cytosine within the CCWGG motif, where W

---

5 Xi and Li. 2009. 'BSMAP: Whole genome bisulfite sequence MAPping program.' *BMC Bioinformatics* 10: 232.

6 Chen P. Y., Cokus S. J. and Pellegrini M. 2010. 'BS Seeker: Precise mapping for bisulfite sequencing.' *BMC Bioinformatics* 11: 203.

7 Krueger and Andrews. 2011. 'Bismark: A flexible aligner and methylation caller for Bisulfite-Seq applications.' *Bioinformatics* 27: 1571–1572.

8 Kahramanoglou C., Prieto A. I., Khedkar S., Haase B., Gupta A., Benes V., Fraser G. M., Luscombe N. M. and Seshasayee A. S. 2012. 'Genomics of DNA cytosine methylation in *Escherichia coli* reveals its role in stationary phase transcription.' *Nature Communications* 3: 886.

stands for an A or a T. Consistent with this knowledge, methylation was detected only within this motif. In addition, the researchers performed the bisulphite sequencing experiment for a Dcm^- strain. Little methylation was detected here, and the very low percentage of methylated bases in this strain was consistent with the stated ability of the bisulphite reagent to achieve ~99% conversion. These reiterated that Dcm was the only DNA cytosine methyltransferase in this *E. coli* strain.

A comparison of the distribution of R_{met} values across the four phases of growth studied showed that the internal cytosine in almost all the CCWGG motifs was fully methylated in the stationary phase; however, in exponentially growing cells, only a subset of reads mapping to these cytosines supported the presence of methylation, as evidenced by $R_{met} << 1$ for many target sites (Fig. 6.4). To validate this finding by a parallel method, the researchers used the MeDIP approach using an antibody specific to 5-methyl cytosine. Consistent with the previously-described caveat of the MeDIP approach, the researchers first found that higher signals were obtained from regions containing a cluster of closely-spaced CCWGG motifs than those with a single or a few target sites, even though both groups of

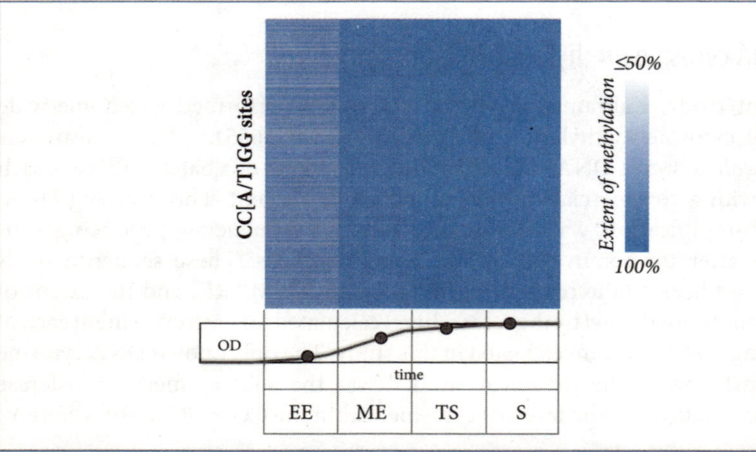

Figure 6.4 *Variation in extent of cytosine methylation with growth phase in E. coli.* This figure shows a heatmap indicating the extent of methylation of each CCWGG motif in the genome of *E. coli* across four different stages of batch growth: Early exponential (EE), mid exponential (ME), transition to stationary (TS) and stationary (S). These were determined by bisulphite sequencing of the genomic DNA, followed by deep sequencing on the Illumina platform. Many CCWGG motifs are only partially methylated in exponential phase, though these appear to be fully methylated during stationary phase.

cytosines may be methylated to the same extent. The researchers corrected for this artifact by performing a non-linear regression of the MeDIP signal against the local density of CCWGG motifs. The authors found that the corrected MeDIP signal was less for those cytosines deemed to be partially methylated from the bisulphite sequencing experiment. It is possible that in rapidly growing cells there is much newly synthesised DNA that is not yet methylated in an average, unsynchronised population. Since the origin of replication fires more than once per cell cycle in rapidly-dividing cells, there is generally more new DNA closer to the origin of replication than further away. If the above-mentioned replication-based model for partial methylation were true, one would expect more partially methylated sites around the origin of replication. However, this was not the case. Instead, the researchers found that partial methylation was distributed across the chromosome, and associated with an extended CCCWGG motif. Thus, it was proposed that this sequence motif could play a role in kinetically slowing down methylation. Next, Kahramanoglou and coworkers discovered that many partially methylated cytosines were present at non-synonymous sites in protein-coding regions. Methyl-cytosine is a mutation hotspot and can rapidly de-aminate to a thymine. Thus, the researchers suggested that partial methylation might be selected to reduce the risk of deleterious non-synonymous mutations in rapidly-dividing cells.

The next question that Kahramanoglou and colleagues set out to answer was the effect of cytosine methylation on gene expression in *E. coli.*[9] Towards this, they performed microarray-based genome-wide gene expression experiments of wild type and Dcm$^-$ *E. coli*. The effects of Dcm$^-$ on gene expression was variable across growth phases, with the most prominent effect seen in stationary phase. An earlier RT-PCR-based study by Militello and co-workers had identified DNA cytosine methylation as a regulator of ribosomal gene expression in the stationary phase.[10] Besides observing this effect again, Kahramanoglou and colleagues noted that Dcm$^-$ significantly up-regulated σS, the stationary phase σ-factor. As a result, many stationary phase-dependent genes were more-than-usually up-regulated in a Dcm$^-$ mutant. This established DNA cytosine methylation as a regulator of the stationary phase gene expression in *E. coli*. However, a relevant phenotype remains to be discovered.

9 The effects of DNA adenine methylation on bacterial gene expression is better established. We do not review these here, but interested readers may refer to the reviews by Casadesus, referred to earlier.

10 Militello K. T., Simon R. D., Qureshi M., Maines R., VanHorne M. L., Hennick S. M., Jayakar S. K. and Pounder S. 2011. 'Conservation of Dcm-mediated cytosine DNA methylation in *Escherichia coli*.' *FEMS Microbiology Letters* 328: 78–85.

6.4 Detecting DNA methylation by single-molecule real-time sequencing

The bisulphite sequencing technique is applicable to the detection of DNA cytosine methylation.[11] Till recently, there were no methods available to detect and quantify DNA adenine methylation at a single-base resolution genome-wide. As mentioned earlier, the resolution of the MeDIP approach is poor and its performance heavily affected by the sensitivity of the antibody. One could always use restriction digestion of the DNA by an enzyme that cleaves only methylated sequences followed by high-throughput end sequencing of the resulting fragments, but this requires prior knowledge of the sequence motif that is methylated and there are concerns about the wide size range of the fragments generated.

The current state-of-the art technique for detecting DNA base modifications, including DNA methylation of various types, is the single-molecule real-time (SMRT) sequencing technology by Pacific Biosciences, which we had only briefly described in Chapter 4. The SMRT sequencing technology uses what are called zero mode waveguides (ZMW), which are small nanoscale volumes capable of capturing a single molecule of DNA, albeit resulting in a high concentration of the DNA within the miniscule volume. The space also serves as a detection volume for fluorescence. DNA polymerase molecules are immobilsed on these ZMWs and the DNA to be sequenced added. Each free nucleotide base, used by the immobilised DNA polymerase to synthesise the complementary strand, is labelled with a different fluorophore. The fluorescence from the base that is being incorporated into the nascent strand is detected only when it is captured inside the ZMW by the immobilised DNA polymerase. Once the base is added, the fluorescent moiety is cleaved off and it diffuses out of the ZMW after which its fluorescence is no longer detected. Thus, during the addition of a new base by the immobilised DNA polymerase, a pulse of fluorescence is detected from the corresponding ZMW–the colour of the fluorescence indicates the base that is added. An array of ZMWs can be placed on a chip, thus enabling multiple parallel sequencing reactions to be performed. This technology is capable of producing ultra-long reads with a median length of 5 kb.

An important parameter that can be measured during an SMRT sequencing experiment is the time gap between the incorporation of two consecutive bases, or the inter-pulse duration (IPD).[12] Researchers at Pacific Biosciences discovered that the presence of modified bases alters the kinetics of DNA synthesis as defined by the IPD, i.e., different base modifications produce different IPD signatures.[13] However, since the IPD signatures are dependent on the local

---

11 5-methyl cytosine and 5-hydroxymethyl cytosine.

12 This is in addition to being able to measure the time duration of the fluorescence pulse itself.

13 (a) Flusberg B. A., Webster D. R., Lee J. H., Travers K. J., Olivares E. C., Clark T. A., Korlach J. and Turner S. W. 2010. 'Direct detection of DNA methylation during single-

sequence context–independent of DNA modification–the comparison with an appropriate null *model*, which accounts for these characteristics, is required. This null model can be a *demodified* DNA of the same sequence as the native DNA, or a computational model which can accurately predict IPD signatures for unmethylated DNA of the same sequence as the native DNA. The IPD ratio, at any given site, between the native DNA and the null model, is predictive of the presence of a modified base. Certain base modifications produce better signals than others. For example, adenine methylation produces the best signals, thus resulting in very high sensitivity as well as specificity (Fig. 6.5). This is believed

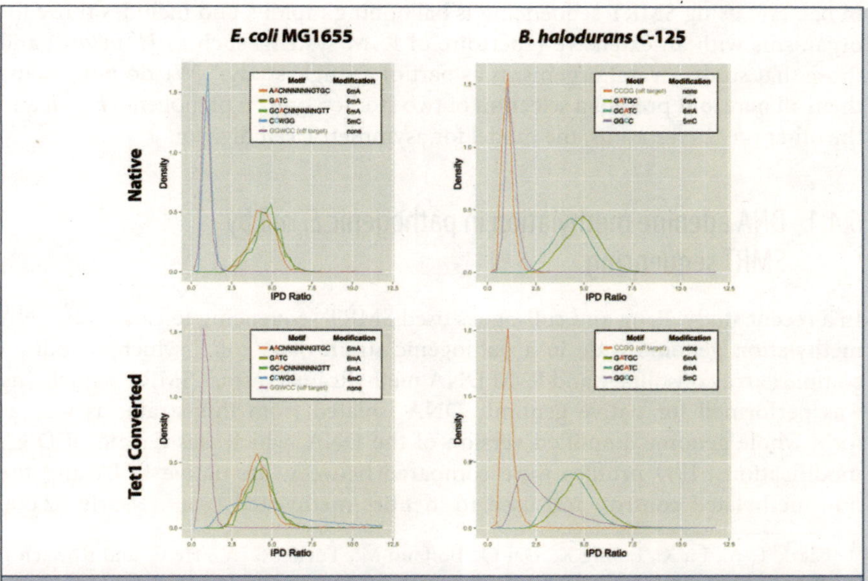

Figure 6.5 *IPD ratio distribution DNA methylation by SMRT sequencing.* This figure shows the IPD ratio distributions for methylated adenines and cytosines in two different bacterial genomes. The IPD ratio distribution of methylated cytosines is low, indicating low sensitivity of the default SMRT sequencing approach to this modification, but is enhanced by subjecting the genomic DNA to Tet1 oxidation. This figure is reproduced under the Creative Commons Attribution Licence from Clark T.A., Lu X, Luong K, Dai Q, Boitano M, Turner SW, He C, Korlach J. 2013. 'Enhanced 5-methylcytosine detection in single-molecule, real-time sequencing via Tet1 oxidation.' *BMC Biol.* Jan 22; 11: 4. © Clark et al. 2013; licencee Biomed Central.

molecule, real-time sequencing.' *Nature Methods* 7: 461–65; (b) Pacific Biosciences White Paper titled, 'Detecting DNA Base Modifications Using Single Molecule, Real-Time Sequencing'.

to be because adenine methylation directly affects the nitrogen atoms involved in base pairing. Cytosine methylation provides weak signals. However, it has been shown that oxidation of the DNA by an enzyme called Tet1, which converts 5-methyl cytosine to the larger 5-carboxyl cytosine, considerably enhances the signal, resulting in highly sensitive detection of cytosine methylation as well.[14] In summary, the SMRT sequencing technology, besides producing ultra-long sequencing reads that can considerably simplify genome and transcriptome assembly, also allows modified bases to be detected as part of a DNA sequencing experiment. This is now becoming a standard method for detecting modified bases on a genomic scale. The number of studies cataloguing DNA methylation in bacteria using SMRT sequencing is ballooning rapidly, and includes those for organisms with an extensive repertoire of R–M systems such as *H. pylori*[15] and those that study several organisms as part of a single study.[16] We do not discuss them all here, but present a selection of two studies, one on pathogenic *E. coli* and the other on *C. crescentus*, the model for asymmetric cell division.

6.4.1 DNA adenine methylation in pathogenic *E. coli* by SMRT sequencing

In a recent study, Fang and colleagues used SMRT sequencing to catalogue DNA methylation genome-wide in a pathogenic strain of *E. coli*,[17] which encodes a complex array of solitary and R–M DNA methyltransferases.[18] SMRT sequencing was performed for native genomic DNA isolated from this strain, as well as for a whole genome amplified version of the DNA, which was devoid of DNA modifications. IPD profiles were compared between the native DNA and the non-methylated control, and used to identify methylated bases. Nearly 52,000

[14] Clark T. A., Lu X., Luong K., Dai Q., Boitano M., Turner S. W., He C. and Korlach J. 2013. 'Enhanced 5-methylcytosine detection in singlemolecule, real-time sequencing via Tet1 oxidation.' *BMC Biology* 11: 4.

[15] Krebes J., Morgan R. D., Bunk B., Spröer C., Luong K., Parusel R., Anton B. P., König C., Josenhans C., Overmann J., Roberts R. J., Korlach J. and Suerbaum S. 2013. 'The complex methylome of the human gastric pathogen Helicobacter pylori.' *Nucleic Acids Research*.

[16] Murray I. A., Clark T. A., Morgan R. D., Boitano M., Anton B. P., Luong K., Fomenkov A., Turner S. W., Korlach J. and Roberts R. J. 2012. 'The methylomes of six bacteria.' *Nucleic Acids Research* 40: 11450–11462.

[17] Serotype O104: H4, causative agent of hemolytic uremic syndrome.

[18] Fang G., Munera D., Friedman D. I., Mandlik A., Chao M. C., Banerjee O., Feng Z., Losic B., Mahajan M. C., Jabado O. J., Deikus G., Clark T. A., Luong K., Murray I. A., Davis B. M., Keren-Paz A., Chess A., Roberts R. J., Korlach J., Turner S. W., Kumar V., Waldor M. K. and Schadt E. E. 2012. 'Genome-wide mapping of methylated adenine residues in pathogenic *Escherichia coli* using single-molecule real-time sequencing.' *Nature Biotechnology* 30: 1232–1239.

modified base positions (genome size of ~10 million bases across the two strands of DNA) were found to show IPD characteristics indicative of DNA methylation. Nearly 95% of these were adenine and 3% were cytosines; the remaining 2% mapped to thymines and guanines. Fang and co-workers performed a motif analysis around the methylated adenines and identified the following sequences: G^{m}ATC, CTG^{m}CAG and ACC^{m}ACC. Nearly 95% of these sites in the genome were detected as methylated. In contrast, only 0.3% of cytosines in the CCWGG motif were found to be methylated, despite the presence of the appropriate methyltransferase; this reflects the poor signal that methylated cytosines produce in an SMRT experiment. A second round of gapped-motif detection around methylated sites not covered by any of the above-mentioned motifs revealed the presence of more complex target sequences for adenine methylation. That these were bonafide targets was suggested by the finding that ~95% of these sites in the genome were methylated. Together, these show the high sensitivity and specificity of SMRT sequencing for unbiased, genome-wide detection of adenine methylation.

Next, the researchers set out to map each of the methylated sequence motifs to a DNA methyltransferase encoded in the genome. For this, they expressed the gene encoding each DNA methyltransferase from a plasmid in a strain of *E. coli* that was totally devoid of DNA methyltransferases. They then performed SMRT sequencing of plasmid DNA isolated from this bacterium, thus enabling direct mapping of methylated sequences to the expressed methyltransferase. In this manner, ~95% of all methylated sequences–including the above-mentioned complex motifs–could be mapped to a methyltransferase. Nearly half of the remaining sites were found adjacent to a bonafide methylated site; this was interpreted as a consequence of a DNA methylation event affecting IPD signatures of other, proximal base positions. Many of the remaining sites could potentially be off-target effects occuring at sequences that closely resemble the methylation target.

Finally, Fang et al. investigated the phenotypic effects of a prophage-encoded R–M system in this *E. coli.* They found that the expression levels of ~40% of all genes are directly or indirectly affected by this system. Deletion of this R–M system adversely impacted growth. This, coupled with the evidence that artificial introduction of the R–M system in a different *E. coli* strain negatively affected the fitness of the recipient, suggested that the native host of this R–M system was uniquely adapted to it.

6.4.2 Insight into the epigenetic control of *Caulobacter crescentus* cell cycle from SMRT sequencing

C. crescentus is an exciting model for studying DNA methylation, as the solitary DNA adenine methyltransferase CcrM is a master regulator of the organism's cell cycle. Further, the expression of CcrM is in turn dependent on the stage of the cell cycle: It is expressed towards the end of the cycle and then proteolytically degraded rapidly. The target site of CcrM–GANTC–has a restricted occurrence (~40% of random expectation); moreover ~25% of these sites are located in intergenic regions, which comprise only ~9% of the *C. crescentus* genome. This suggests a possible regulatory role for GANTC methylation in this organism.

Kozdon and colleagues performed SMRT sequencing of *C. crescentus* genomic DNA isolated at different stages of the cell cycle.[19] The authors found that two other motifs, in addition to GANTC, are methylated at adenines in this organism. In contrast to GANTC, these motifs are not particularly enriched in intergenic regions. Since these are targeted by R–M systems, which may be parasitic and recently acquired, they may not have a selective value to the organism. In contrast to the above-described study of methylation in *E. coli*, Kozdon and co-workers performed Tet1 oxidation of the DNA before searching for DNA cytosine methylation, and therefore discovered that 50–75% of two motifs are cytosine methylated in this organism.

Following the discovery of novel motifs for DNA methylation, Kozdon and researchers focused on the dynamics of GANTC methylation during the cell cycle. Immediately after the initiation of DNA replication, CcrM is not present in the cell. Therefore, at this point newly synthesised DNA close to the origin remains hemi-methylated. Similarly, over the course of the cell cycle, newly synthesised DNA remains hemi-methylated, until the late expression of CcrM results in its (near-)complete methylation. Consistent with this biology, Kozdon et al. noticed that IPD signals for DNA closer to the origin of replication were lower than those near the terminus for most of the cell cycle, and the region of transition from low to high IPD values should define the location of the progressing replication fork. In contrast to the GANTC motif, the other target sites for DNA methylation were found to be fully methylated immediately after their replication, indicating the absence of cell-cycle control over the activities of the corresponding methyltransferases.

Kozdon et al. identified ~25 GANTC sites that were persistently unmethylated across the cell cycle. In particular, they noticed that three intergenic regions, which contained a few of these motifs, were annotated as 'essential' because they could not be disrupted by insertion mutagenesis. However, these could be deleted by other mechanisms, and the resulting mutants were viable. This suggested the existence of a protection mechanism that was responsible for both the failure of insertion mutagenesis as well as the persistently unmethylated state of these sites.

---

[19] Kozdon J. B., Melfi M. D., Luong K., Clark T. A., Boitano M., Wang S., Zhou B., Gonzalez D., Collier J., Turner S. W., Korlach J., Shapiro L. and McAdams H. H. 2013. 'Global methylation state at base-pair resolution of the Caulobacter genome throughout the cell cycle.' *Proceedings of the National Academy of Sciences USA* 26: 110, 48: E4658–67.

Finally, the researchers sought to identify candidate genes for cell-cycle-dependent epigenetic transcriptional regulation. From other genome-scale, experimental annotation studies which identified transcription start sites (TSS) for all *C. crescentus* genes, the authors defined ~110 TSS as differentially expressed across the cell cycle as well as containing at least one GANTC motif. A majority of these sites were found to become hemi-methylated in a time-window that coincided with a cell-cycle-dependent change in their expression levels. Thus, this study identified ~60 genes as candidates for epigenetic gene expression control in *C. crescentus*.

Summary

- ✓ DNA methylation is an epigenetic mark, not only in higher eukaryotes, but also in bacteria.
- ✓ Many DNA methylating enzymes probably evolved from horizontally acquired ancestors in the form of restriction–modification systems.
- ✓ Immunoprecipitation and bisulphite sequencing are two manipulation techniques that help identify sites of DNA methylation in bacteria.
- ✓ Single-molecule real-time sequencing enables methylation calls as part of a normal genome sequencing experiment.

Index